*Protection Against
Atmospheric Corrosion*

Protection Against Atmospheric Corrosion

Theories and Methods

KAREL BARTOŇ

Translation:

John R. Duncan,

Nottingham University

A Wiley–Interscience Publication

JOHN WILEY & SONS

London · New York · Sydney · Toronto

First published 1973 © Verlag Chemie GmbH, Weinheim/Bergstr.
under title *Schutz gegen atmosphärische Korrosion—Theorie und
Technik* by Karel Bartoň.

Copyright © 1976, by John Wiley & Sons, Ltd.

Library of Congress Cataloging in Publication Data:

Bartoň, Karel.
  Protection against atmospheric corrosion.
  Translation of Schutz gegen atmosphärische Korrosion.
  'A Wiley–Interscience publication.'
  1. Corrosion and anti-corrosives. I. Title.
TA462.B37313   620.1'1223   75-26570
ISBN 0 471 01349 8

Photosetting in Great Britain by Technical Filmsetters Europe Limited, 76 Great Bridgewater Street,
Manchester M1 5JY and printed by The Pitman Press Ltd., Bath, Avon.

Preface

This monograph attempts to summarize those particular sections of corrosion science and technology involved in atmospheric corrosion. It is my aim in writing it that it should bridge the gap between theory and practice, which has become very marked in this field.

Atmospheric corrosion is a typical interdisciplinary field, drawing on theoretical and practical knowledge from many different sciences (metallurgy, physical and macromolecular chemistry, engineering, and meteorology, among others). Only by recognizing this can the protection of metals against atmospheric corrosion be moved from its present empirical state and put on a theoretical footing. Accordingly, a review of the theoretical aspects is followed by chapters in which I have tried to describe the thoughts of a corrosion engineer seeking to apply these basic principles. Comprehensive descriptions and discussions of technical data, such as standards, specifications, etc., have been avoided.

The reader may judge how successfully I have achieved this aim. I should be thankful for any critical comments.

I wish to thank my colleagues at the G. V. Akimov State Institute for Materials Protection for helpful discussions. Without the help of my wife in the planning and production of the manuscript, this book would not exist. It is therefore dedicated to her.

1706 Bělohorská,
Praha 6, CSSR
September, 1972

KAREL BARTOŇ

Contents

1

Introduction

1.1 The technical and economic significance of atmospheric corrosion

As was pointed out by Uhlig [1], corrosion protection and the damage which is caused by corrosion are important burdens on the economic life of the society. Approximately half the total cost of corrosion protection in the USA in 1949 (over $US2.8M) was spent on measures against atmospheric corrosion effects. The cost of paint, which protects against corrosion under most atmospheric conditions, was 37.67%. Phosphating (0·38%), galvanizing (2·52%), and parts of the use of nickel (approximately 2·5%) and cadmium plating (0·39%) also belong in this category. The sum of these measures against atmospheric corrosion exceeds 43%. There must also be taken into account a certain part of the use of non-rusting steels and other corrosion-resistant alloys and metals (e.g. alloys of copper and aluminium). Thus, it seems reasonable to assume that about 50% of the total cost of corrosion protection is spent on measures against atmospheric corrosion.

Some further data show convincingly the economic importance of protection against atmospheric corrosion: Daeves and Trapp [2] calculated in 1937 that the annual conversion of iron into rust during atmospheric corrosion accounted for approximately 2% of the current German steel production. Since these calculations were based on reliable estimates of the total unprotected iron surface and average corrosion rates, they should still be valid today, when the corrosion rates will be even higher, due to increased air pollution. Direct conversion to rust thus accounts for a loss of 20,000 tons from each million tons of steel produced annually. The value of this amount of steel may not be especially great, but it should be remembered that steel products lose their usefulness long before the steel is completely converted to rust. Motor vehicles are a widely-known example of this effect.

Most surfaces which are endangered by atmospheric influences are protected in various ways (by paint, metallic coatings, etc.). The costs of these protection measures are extraordinarily high. They may be estimated relatively accurately from statistical data on the use of protective materials (i.e. paint products, metals, chemicals, etc.). The data in Table 1 were calculated for costs, mostly of protection against atmospheric effects, in Czechoslovakia in 1967.

Considerable damage is caused by insufficient temporary protection. Corrosion effects on working surfaces of machinery, deterioration of properties

1

Table 1.

	10^6 Kčs
New production (i.e. Machinery (including motor vehicles), surface-protected manufactured articles, and building materials)	2 725
Maintenance of corrosion protection on existing products	1 950
	4 675

of electrical equipment, surface disfiguration, etc., which often occur during transport, storage, and temporary exposure out of doors, are in themselves hardly economically significant, but should not be underrated.

Many factors must be taken into account in making an assessment of the economic importance of atmospheric corrosion: the unnecessary labour to manufacture products and protective coatings destroyed prematurely by corrosion (including the production of paints and other corrosion protection agents); losses during corrosion protection maintenance because of equipment shut-down (e.g. high-voltage lines), etc.

It is therefore not surprising that basic research is going on all over the world into the extremely complex problems of atmospheric corrosion, and the development of new, effective protection methods.

1.2 The meaning of the concept of atmospheric corrosion

Aggressive action by the atmosphere is the widest-known form of corrosion. The atmospheric rusting of iron, in particular, was described in the earliest natural histories. Thus Pliny the Elder mentions that rust arises by simultaneous action of water and air. The Phlogiston Era of chemistry dealt with this phenomenon, and explained it by the reaction Iron = Rust + Phlogiston.

After the discovery of the electrochemical basis of corrosion, it was soon recognized that the effect of the atmosphere on metals was due to this phenomenon. Similarly, it was found that the general laws which hold for metal corrosion in electrolytes also hold for the special case of atmospheric corrosion.

The problems involved in corrosion of metals in their active states, especially with formation of solid corrosion products, are far from solved. This is in contrast to the exactly-formulated description of the reaction processes in high-temperature corrosion (scaling) of metals and simple alloys in oxidizing gases [3], and to the partly explained, though not uniformly interpreted, laws of metal passivation and corrosion kinetics of passive metals. As will be shown later, atmospheric corrosion is a boundary case between active dissolution of metals and corrosion in the passive region.

It has been shown by the work of Evans [4], Tomashov [5], and Kaesche [6], in particular, that electrochemical corrosion can generally be interpreted as the result of a series of alternate chemical and physical processes. The kinetically-controlling (i.e. slowest) reaction can be determined, but in many cases it is difficult to understand the quantitative kinetics of the individual processes. It is often difficult to decide which part-processes may be neglected; the kinetics of the overall reaction change with time, due to a changing rate of the controlling reaction, or to transfer to another rate-determining step. (For example, a chemically-controlled process may become transport-controlled following formation of a reaction layer).

There is a special problem in the case of atmospheric corrosion which makes an exact description of the reaction processes, and especially the kinetics of the electrochemical corrosion, difficult: atmospheric corrosion involves the existence of a limited amount of electrolyte, and the formation, properties and destruction of this electrolyte layer are influenced by a wide range of different factors. Among these are the chemical composition and physical properties of the atmosphere, and, in particular, the solid corrosion products formed during the corrosion processes. Very little is known of the effective microclimate at the surface; this is strongly affected by the heat capacity of the corroding object, and the way in which it is built. The variability of the atmosphere, which is related mainly to geographical effects and the influence of Man, is the most important factor.

The nature of atmospheric corrosion can be defined as follows: atmospheric corrosion is an electrochemical process which occurs in a limited amount of electrolyte. The electrolyte is neutral or slightly acidic (or, under exceptional conditions, slightly alkaline), and its properties are influenced chiefly by the chemical composition of the atmosphere and the properties of the corrosion products formed. The neutral or slightly acidic nature of the electrolyte and its variable presence on the corroding surface promote the formation of solid corrosion products on all metals which remain unpassivated for thermodynamic reasons. This is easily understood, since the solubility product of the reaction product is easily exceeded in the small volume of approximately neutral electrolyte, and so new phases are formed in the system [7]. The properties of this reaction layer are not constant, however, since the temporary presence of the electrolyte layer (as was discussed earlier) plays an important role.

Crystallization processes in the corrosion product layer can change its properties significantly. Periodic renewal of the electrolyte can wash the soluble component of the corrosion product out of the layer, especially if it occurs intensively (e.g. during rain).

1.3 Important unanswered questions regarding atmospheric corrosion

As was discussed earlier, economic considerations are the force behind the intensive attempts to understand atmospheric corrosion and to develop effective protection methods. As a result of the considerable research into atmospheric corrosion and protection, the most complete possible knowledge of the laws governing these processes and an understanding of the mechanisms of action of the different protection methods should be obtained.

The aim of this monograph is to provide a critical summary of the basic knowledge on these subjects, and so derive principles of optimum protection measures.

Though atmospheric corrosion is very widespread and not known only to experts, a thorough analysis of the knowledge in this field shows that most protection measures rely on empirical experience, and are only slightly influenced by theory. Without doubt, a greater application of such theoretical ideas would bring good economic side effects.

Further, if a series of basic data remains to be collected, it should be possible to apply theoretical ideas in devising systems from which this data could be obtained. The most important problems to which this might be applied are:

1. Comprehension of the long-term course of atmospheric corrosion in terms of the composition of the metal and the physical and chemical properties of the atmosphere, and their changes due to geographical and human influences (pollution of the air, which influences meso- and micro-climates, etc.). In this connection, it appears necessary to define the properties of the specific micro-climates at the corroding surfaces, since they arise differently in different atmospheres due to formation and mode of action of the particular products. It is thus necessary to interpret the long-term course of the process as the sum of short-term processes, whose kinetics should be more easily understood.

2. A practical result of this knowledge could be a firmly based classification of atmospheric corrosion action on metals, which might serve as a basis for standards of finish on different metal products (such as structures or machinery) whose lifetime and reliability of performance are impaired in different ways by atmospheric corrosion. Such a classification would be very useful; with its help, technically and scientifically optimized protection measures could be selected and implemented for particular atmospheres which were well-defined from a corrosion viewpoint. This should ensure reliability of the product (as far as it is affected by corrosion) throughout its lifetime, and simultaneously allow a much reduced cost of upkeep.

3. Closely connected to the first two points is the problem of test techniques for evaluation of the lives of unprotected metallic materials or those with coatings. Since atmospheric corrosion is a relatively slow process, it is important to use methods which will allow pre-determination (or at least estimation) of the long-term course of the process using short-period testing. This can only be achieved by a knowledge of the kinetics of the natural process, which should be the basis for theoretically-formulated modelling or short-term tests, and corresponding evaluation methods.

4. Coatings are used very widely for protection against atmospheric corrosion. Their lives and associated usefulness are, in general, too low, due chiefly to the rather empirical methods of selection and application. Little is known about the corrosion mechanism in the metal-protective coating system in different atmospheres, and a still smaller part of this knowledge is put into practice. An exact understanding of the corrosion behaviour of metal-protective coating systems, and its practical application in developing new protection systems, would be of considerable benefit.

2

The Atmosphere as a Corrosive Environment

2.1 Definition of the concept of 'atmosphere'

The atmosphere, and especially its chemically active components, causes the atmospheric corrosion of metals. It thus seems sensible to consider the chemical and physical properties of the atmosphere.

Since atmospheric corrosion proceeds by an electrochemical mechanism, special attention will be paid to properties which affect this process.

2.2 The chemical composition of the atmosphere

2.2.1 Nitrogen and the inert gases

Nitrogen and inert gases can be ignored in corrosion, since although they form the majority of the atmosphere, they do not react with metal surfaces in normal environments.

2.2.2 Oxygen and ozone

In marked contrast, oxygen is certainly a very important atmospheric component and has an intensive influence on the corrosion mechanism because of its high partial pressure (around one fifth of total air pressure) and its high reactivity. Atmospheric corrosion is an oxidation-reduction process, in which the metal participates as an electron donor and oxygen, and other reducible species, as electron acceptors.

Ozone, which occurs in concentrations of approximately 10 to 50 μg m^{-3}, is particularly active in promoting aging of elastomers and other organic substances used as coatings.

2.2.3 Water

Water is the third greatest component of the atmosphere by concentration. In contrast to the two major components of the air, its concentration in the air is not constant, but varies between wide limits. Further, water occurs in all its states: in solid form as snow, ice and hoar-frost, liquid as rain, fog, dew, or thawing ice or snow, and as gaseous water vapour. The water content of the

5

6

atmosphere must be discussed in terms of precipitation and absolute humidity (atmospheric vapour content). Annual precipitation may exceed 3000 mm (especially in tropical rain forest climates), and is influenced mainly by geography [1].

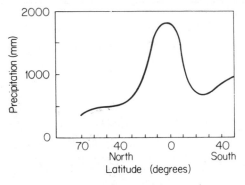

Figure 1. Precipitation on the continents at different latitudes.

The water vapour content (absolute humidity) of the atmosphere occasionally climbs above $30 \, g \, m^{-3}$ (approximately 31 Torr) in warm, damp regions, and like precipitation is dependent on geographical position [2].

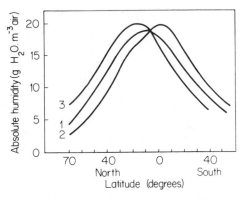

Figure 2. Absolute atmospheric humidity at different latitudes.
1. Annual average value; 2. Average value for December to February; 3. Average value for June to August.

Atmospheric water undoubtedly plays a decisive role in atmospheric corrosion. The presence of liquid water on the surface is an important prerequisite for the electrochemical path of the corrosion process. Since liquid water is deposited on the surface not only directly by precipitation (rain, fog, dew, thawing snow) but also by adsorption and condensation processes, which are determined jointly by the absolute humidity and the temperature, the concept of relative humidity is very important in atmospheric corrosion. The relative humidity is the ratio of the absolute humidity to the saturation value, and is expressed as a percentage.

Liquid water in the atmosphere dissolves components of the solid matter and gases present in the atmosphere, so that it represents a solution of different species. Its pH value can fall as low as 3. It is always saturated with oxygen. If sulphur dioxide or other gases are present in the air, it also absorbs these gases. It is obvious that precipitation occurring as fine drops (i.e. mist or fog) will absorb particularly large amounts of aggressive species from the air because of its large surface area.

Long periods of rain promote corrosion by preserving the surface electrolyte necessary for the corrosion, but may also inhibit corrosion during the subsequent period because of partial washing-away of soluble stimulating species from the corroding surface.

Liquid water on the surface also plays an important role in the disruption of protective organic coatings by atmospheric effects.

2.2.4 Carbon dioxide

The concentration of this component of the atmosphere, which is mostly naturally-occurring, is usually of the order of $10^{-2}\%$. The highest values, which may be a few tenths of a percent, occur only exceptionally, in special enclosed microclimates such as caves or stables. While this relatively high concentration and the appreciable solubility of the gas in water (0.232 g CO_2 per 100 g H_2O at $10\,°C$) could be a theory to account for part of the corrosion, carbon dioxide appears unimportant in atmospheric corrosion. It does, however, participate in formation of secondary products, especially on non-ferrous metals. Kaesche [3] has given an interesting theoretical discussion of this point, which might not be expected at first sight. He concludes that dissolution of atmospheric carbon dioxide will give an electrolyte pH of between 5 and 5·6. If oxygen is present in excess, there is no transition at this pH from oxygen reduction to H^+ reduction as the cathodic partial process. Far higher partial currents can be obtained from oxygen reduction than from hydrogen evolution at these carbon dioxide concentrations. Carbon monoxide appears not to affect the course of atmospheric corrosion.

2.2.5 Sulphur dioxide

This is the typical gas impurity found in urban and industrial atmospheres. It is emitted to the atmosphere in large amounts during combustion of sulphur-containing fuels of all types, so that the sulphur dioxide content of the air may exceed several mg m^{-3}. Expressed as a percentage, this means concentrations in polluted atmospheres of 10^{-6} to $5 \times 10^{-4}\%$ (i.e. 0·01 to 5 parts per million (ppm)) [4]. Sulphur dioxide is an extremely reactive gas, and is the most important atmospheric corrosion stimulant because of its high solubility in water (16·2 g SO_2 per 100 g H_2O). (See Chapter 3.5).

2.2.6 Hydrogen sulphide

Hydrogen sulphide is another pollutant with significant effects on corrosion, despite its concentration only rarely exceeding 100 μg m^{-3} i.e. 0·06 ppm. This gas promotes the corrosion of copper, nickel and silver, in particular; i.e. of those metals with high energies of sulphide formation.

2.2.7 Other acidic gases

Nitric oxide occurs in urban atmospheres at concentrations up to approximately 1 μg m^{-3}, though higher values (10 to 20 μg m^{-3}) may occur around nitric acid factories, etc. Nitric oxide is formed mainly in thunderstorms [5], and hardly affects atmospheric corrosion.

Chlorine and hydrogen chloride are also frequently found in the air near chemical works, etc. There is no accurate data on their concentrations, though these species are extremely aggressive.

Formaldehyde, formic acid and acetic acid are similarly powerful stimulators of atmospheric corrosion. Their presence in packages and cases usually stems from degradation of organic packing materials (glue, plastics, wood, etc.) [6].

2.2.8 Salts

Chloride is a chemically active component of the atmosphere, especially in coastal areas. Its concentration falls with distance from the coast, as shown in Figure 3. Highest chloride ion values are found in the breaking waves (up to 3 mg m^{-3}). Average values near to the sea are an order of magnitude smaller

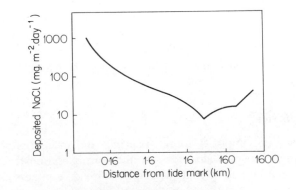

Figure 3. Deposition of sodium chloride aerosol as a function of distance from the tide mark.

(approximately $100 \ \mu g \ m^{-3}$ at maximum). Maximum values in industrial and urban atmospheres are around 10 to $30 \ \mu g \ Cl^- \ m^{-3}$.

Other corrosion-stimulating salts (e.g. ammonium sulphate or nitrate, or others) can occur from time to time under special conditions.

There is a pronounced corrosive effect from salts used (in large quantities) in winter as anti-icing agents.

2.2.9 Ammonia

The ammonia content of the air is seldom higher than $50 \ \mu g \ m^{-3}$ [4]. Ammonium salts have been identified recently in atmospheric corrosion products (especially rust), and the ammonia in the atmosphere could play some part in the corrosion process. However, Ross and Callaghan [7] suggest that the ammonium salts are formed chiefly from the nitrogen content of the corroding material (steel). Ammonium salt solutions have a low surface tension, and thus cause sheet- rather than drop-forming condensation on surfaces liable to corrode, so that the corrosion can proceed all over the surface.

2.2.10 Dust

The amount of dust in the atmosphere, and its composition, are very variable, and depend on many factors. The nature of the earth's surface and air pollution from various industrial works play decisive roles. In rural regions, the atmospheric dust is composed mainly of organic and inorganic components of the earth's surface, whereas in urban and industrial regions there is typically a higher content of industrial pollutants. The water-soluble inorganic component of these dusts is also generally higher. The absolute quantity of dust in the atmosphere varies between less than 10^3 to more than 2×10^5 dust particles per cubic centimeter. Kutzelnigg [4] quotes $2 \cdot 3 \ mg \ m^{-3}$ as a normal value for urban air, and over $200 \ mg \ m^{-3}$ in specific environments e.g. around cement works. The size (diameter) of the dust particles lies between approximately $0 \cdot 1 \ \mu m$ and $0 \cdot 3 \ mm$ [8].

2.2.11 Collected data on air impurities

Kutzelnigg [4] has produced a review of the concentrations of the individual air impurities (Table 2, Figure 4). Rosenfeld [9] gives the following analysis of dust in an industrial district in Moscow:

Total matter (soluble and insoluble)	$91 \ g \ m^{-2}$ surface.
Inorganic component	69
Organic component	33
Sulphate	0·35
Chloride	27

Bartoň [10] reports values of up to 3% soluble sulphate in industrial dust.

Table 2. The occurrence of atmospheric impurities (from Kutzelnigg [4])

Species	'Normal values' for urban atmospheres	Limiting values
CO_2	0·03%	0·306% (caves) 0·275% (cow shed)
CO	12·5 mg m^{-3}	29 mg m^{-3}
pH of precipitation	5·5	2·9 to 7·2
Dust	2·3 mg m^{-3}	224 mg m^{-3} (cement factory)
SO_2	0·11 to 2·3 mg m^{-3}	10 to 50 mg m^{-3} (industrial area)
Cl^-	32 μg m^{-3}	3 mg m^{-3} (breaking surf)
O_3	15 μg m^{-3}	55 μg m^{-3}
NH_3	30 to 60 μg m^{-3}	675 μg m^{-3}
NO_2	0·2 to 2·5 μg m^{-3}	23 μg m^{-3}
H_2S	1 μg m^{-3}	0·1 to 10 mg m^{-3}
Formaldehyde	5 to 50 μg m^{-3}	

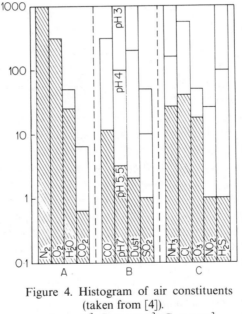

Figure 4. Histogram of air constituents
(taken from [4]).
A. g m^{-3}; B. mg m^{-3}; C. μg m^{-3}.

2.3 Physical properties of the atmosphere important in corrosion

2.3.1 Temperature and its fluctuation

The differences between the separate climatic zones include temperature extremes of $-84\,°C$ and $+58\,°C$. These extreme temperatures are of little importance in atmospheric corrosion, however; at low temperatures (below -5 to $0\,°C$), water usually occurs only in solid form and cannot act as an electrolyte. The highest temperatures occur only in tropical and sub-tropical desert regions, where the relative humidity is too low for electrolyte layer formation. As a general rule, electrochemical atmospheric corrosion proceeds only rarely at temperatures above 25 °C.

Figures 5 and 6 demonstrate this point. Comparison of plots of relative humidity (using the coordinates month and hour of day) with the corresponding temperature plots shows that at high temperatures the relative humidity is always low, and atmospheric corrosion is thus unlikely.

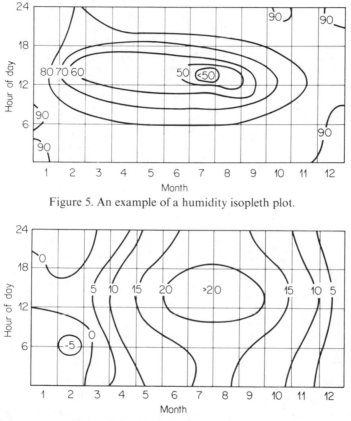

Figure 5. An example of a humidity isopleth plot.

Figure 6. An example of an isotherm plot.

Since atmospheric corrosion is an electrochemical process, it might be expected that it should proceed more rapidly at higher temperatures. This hypothesis does not hold in practice, as will be discussed more closely in Chapter 3.7.2.

Temperature changes bring with them simultaneous changes in relative humidity. At high humidity, a rapid small temperature drop causes exceeding of the dew-point, and so promotes corrosion. A rapid intensive heating of the air has the opposite effect. Since atmospheric corrosion does not require that

Table 3. Absolute atmospheric humidities (expressed as g water vapour $^{m-3}$) at different temperatures and different relative humidities

Temp-erature °C	Relative humidity (%)									
	10	20	30	40	50	60	70	80	90	100
0	0·49	0·98	1·47	1·96	2·45	2·94	3·43	3·92	4·4	4·9
1	0·52	1·04	1·56	2·08	2·60	3·12	3·64	4·16	4·7	5·2
2	0·56	1·12	1·68	2·24	2·80	3·36	2·92	4·48	5·0	5·6
3	0·60	1·20	1·80	2·40	3·00	3·60	4·20	4·80	5·4	6·0
4	0·64	1·28	1·91	2·56	3·20	3·84	4·48	5·12	5·8	6·4
5	0·68	1·36	2·04	2·72	3·40	4·08	4·76	5·44	6·1	6·8
6	0·73	1·46	2·19	2·92	3·63	4·38	5·11	5·84	6·6	7·3
7	0·77	1·54	2·31	3·08	3·85	4·62	5·39	6·16	6·9	7·7
8	0·83	1·66	2·49	3·32	4·15	4·98	5·81	6·64	7·5	8·3
9	0·88	1·76	2·64	3·52	4·40	5·28	6·16	7·04	7·9	8·8
10	0·94	1·87	2·82	3·76	4·70	5·64	6·58	7·52	8·5	9·4
11	0·99	1·99	2·98	3·98	4·97	5·97	6·96	7·96	8·9	9·9
12	1·06	2·12	3·18	4·24	5·30	6·36	7·42	8·48	9·5	10·6
13	1·13	2·26	3·39	4·52	5·65	6·78	7·91	9·04	10·2	11·3
14	1·20	2·40	3·60	4·80	6·00	7·20	8·40	9·60	10·8	12·0
15	1·28	2·56	3·84	5·12	6·40	7·68	8·96	10·20	11·5	12·8
16	1·35	2·72	4·08	5·44	6·80	8·16	9·52	10·90	12·2	13·6
17	1·45	2·89	4·33	5·78	7·22	8·67	10·10	11·60	13·0	14·5
18	1·54	3·07	4·61	6·14	7·68	9·22	10·80	12·30	13·8	15·4
19	1·63	3·25	4·88	6·51	8·13	9·76	11·40	13·00	14·6	16·3
20	1·72	3·44	5·16	6·88	8·60	10·30	12·00	13·80	15·5	17·2
21	1·82	3·65	5·48	7·30	9·13	11·00	12·80	14·60	16·4	18·2
22	1·93	3·87	5·80	7·44	9·67	11·60	13·50	15·50	17·4	19·3
23	2·05	4·10	6·15	8·20	10·25	12·30	14·30	16·40	18·4	20·5
24	2·17	4·34	6·51	8·68	10·85	13·00	15·20	17·40	19·5	21·7
25	2·29	4·58	6·87	9·16	11·45	13·70	16·00	18·30	20·6	22·9
26	2·42	4·84	7·26	9·68	12·10	14·00	16·90	19·40	21·8	24·2
27	2·56	5·12	7·68	10·25	12·80	15·40	17·90	20·50	23·0	25·6
28	2·71	5·42	8·15	10·85	13·50	16·30	19·00	21·70	24·4	27·5
29	2·86	5·72	8·58	11·44	14·30	17·20	20·00	22·90	25·7	28·6
30	3·02	6·04	9·05	12·10	15·10	18·10	21·10	24·10	27·2	30·2

the dew-point be exceeded, but still proceeds at lower humidity values (see Chapter 3.3), and since the absolute humidity is far less variable (Table 3), temperature changes are of special significance in atmospheric corrosion.

2.3.2 Air pressure

The air pressure has no effect on atmospheric corrosion.

2.3.3 Air movement

Wind can influence atmospheric corrosion under some conditions. It often causes rapid temperature and humidity fluctuations and has a considerable influence on the deposition of corrosion stimulators onto the metal surface.

In coastal areas, in particular, the wind direction is of importance in determining the concentration profile of salt particles. Similarly, in industrial regions, wind promotes the transport of corrosion-stimulating air impurities to the surface. Abrasion by wind-borne solid particles can, under exceptional conditions, promote corrosion by disrupting the surface protective layer or coating.

2.3.1 Thunderstorms

Disregarding the usual correlation between thunderstorms and sudden temperature and humidity changes, these storms do not affect the course of atmospheric corrosion. It is doubtful whether the small quantities of nitric oxide, ammonia, etc. formed during thunderstorms can promote corrosion [5].

2.3.5 Radiation

It is unlikely that there should be a significant influence of sunlight, with its different wavelengths, on the atmospheric corrosion of metals which are not covered by protective organic coatings. Some reports in the literature suggest that semi-conducting layers of solid corrosion products (e.g. on zinc or copper) may be activated by the energetic UV band of the solar spectrum, and so affect the corrosion, but the phenomenon is hardly significant enough to have any real effect. There is no doubt that solar radiation has a decisive effect on the aging of protective organic coatings, and that the effect is especially injurious if there is simultaneous action from high humidity. This favours corrosion of the protected metal. UV radiation, in particular, promotes photochemical aging processes, but the IR band of the spectrum is not without effect, since it can accelerate thermal aging by raising the surface temperature to 100 °C or more.

2.4 Microclimate

Up to this point, the atmosphere has been described as a corrosive environment from the standpoint of general meteorological and chemical quantities. These quantities are measured by more or less standard methods all over the world, so that a definition of atmospheric corrosion conditions can be given to a first approximation. However, experience shows that objects under the same defined conditions may corrode quite differently. This is the result of micro-climatic effects, which introduce another series of factors besides the general meteorological conditions. Among these are:

The distance of the object from the earth's surface, or heat-conducting or -insulating connections with the earth. It is well-known that higher humidity values are usually found very close to the earth's surface than at greater distances. Geiger [12] quotes as an example that, at sunset, the relative humidity 5 cm above the ground is 100%, but only 55% at a height of 200 cm. A heat-conducting connection to the earth delays the temperature equilibration of the metal surface with the atmospheric surroundings, since the earth acts as a heat sink.

The mass of the metal product must be taken into consideration. Heavy products, with high heat capacities, exhibit considerable hysteresis of their surface temperature during large temperature variations. At increased atmospheric temperatures and unchanged absolute humidity, the surface remains cooler for a longer time than its environment, and thus often remains wetter longer.

Special heating or cooling around corroding objects plays an important part in the formation of microclimates. Radiant heat produced during the functioning of continuously-working electrical equipment, etc., creates a favorable microclimate for prevention of atmospheric corrosion. On the other hand, a continuously under-cooled surface is a dangerous microclimate, as would be predicted from meteorological knowledge.

The influence of the configuration of a product on protection of it against corrosion is well-known [13, 14], and is closely connected with the creation of specific surface climates. Structures which are not conducive to corrosion protection may be subject to long periods of dampening of the surface, crevice corrosion, or macro-cell action. (This problem will be examined more closely in Chapter 7.) The presence of aggressive gaseous degradation products from organic materials (plastics, rubber, adhesives, wood, etc.) in confined areas, together with enclosed water, is also very favorable to formation of extremely aggressive microclimates. A specific example of a damp microclimate set up during construction is the often-observed phenomenon of inside corrosion of galvanized roofs. Here, water evaporates from damp walls, cannot escape from under the roof, and so condenses on its underside, which then corrodes very rapidly due to the continuous presence of the moisture.

The slope of the surface and its geographical orientation also play important roles in corrosion. Vertical surfaces are always attacked more slowly than horizontal ones. This is obviously because of the easier drainage of precipitated water. The surface facing the earth corrodes more rapidly than that facing away. This is clearly connected with the faster drying and the washing-off of aggressive deposits by rain, etc, from the surface facing away from the ground. Geographical orientation leads to e.g. longer periods of damp on north-facing surfaces. The prevailing wind direction also plays a role here.

Other properties also influence the microclimate. Coloration and surface nature, which chiefly determine the light absorption and heat exchange with the atmosphere, should not be neglected.

3

Mechanisms and Kinetics of
Atmospheric Corrosion

3.1 Oxidation mechanisms and kinetics in dry atmospheres at environmental temperatures

Though it was emphasized in both the foregoing chapters that the nature of technically-significant atmospheric corrosion is predominantly electrochemical, it seems appropriate to devote some attention to the special case of the action of dry atmospheres on metals. Though a dry atmosphere is not particularly corrosive at environmental temperatures, it leads to formation of thin oxide layers during the oxidation process which are important for the later course of corrosion under other conditions [1].

It is scarcely necessary to go into details of theories of metal oxidation here, since notable monographs on this subject are available [2]. However, some inferences from the knowledge of oxidation processes have a general validity, and so they will be recapitulated here, briefly.

Oxidation proceeds via three inter-related part-processes:

1. Reaction at the metal oxide-layer interface.
2. Transport processes in the oxide layer.
3. Adsorption and ionization processes at the boundary between oxide layer and atmosphere.

Each of these three part-processes may be rate-determining, under appropriate conditions.

Three-dimensional oxide layers are ionic crystals with vacancies (i.e. n- or p-type semiconductors). The vacancies enable transport of ions and electrons through the oxide layer. There are frequently also coarser structure defects, such as micropores and cracks. Material transport during oxidation can thus be via the following possibilities:

1. Diffusion of ions and electrons through an homogeneous oxide layer which has only lattice defects.
2. Ion transport through micropores.
3. Transport limited to a part of the surface.

For the first case, the driving force of the transport process is the chemical potential gradients for the diffusing species (ions, electrons) and the consequent electric field gradient. If the linear law, which applies only to those metals

15

which do not form a coherent oxide layer, is neglected, the following laws must be considered:

Parabolic:	$y^2 = kt + a$	(3-1)
Reciprocal-logarithmic:	$l/y = A - B \ln t$	(3-2)
Logarithmic:	$y = k_1 \ln(at + 1)$	(3-3)
Cubic:	$y^3 = k_2 t + a_2$	(3-4)

in which y = weight increase minus oxygen uptake, t = time, and a, k, A and B are constants.

The theoretical foundations and derivations of these laws are due to Wagner, Mott and Cabrera, Hauffe, Hoar and Price, Landsberg [3] and others, who have enabled interpretation of experimental results in terms of physical and kinetic hypotheses. The parabolic and cubic laws of oxidation are valid only for higher temperatures insofar as they apply to atmospheric corrosion.

If it is assumed that, as often happens in such cases, there are mechanical faults in the oxide layer (spaces which fill themselves up during the oxidation process) in addition to the lattice faults, there can be added to the above list the asymptotic law (due to Evans [4]):

$$y = k_3(1 - \exp(-k_4 t))$$ (3-5)

The relationship between the different laws and their ranges of validity may be summarized as follows (Evans [5]):

A. The oxidation rate is controlled only by the transport of ions and electrons through the oxide layer, unaffected by the rates of the surface reactions and without change of the effective surface. At higher temperatures, where the species diffuse without influence of the electric field, the parabolic law is valid, while at lower temperatures, at which transport is via a potential gradient, the reciprocal-logarithmic law is valid.

B. The reaction is influenced by other part-processes in addition to diffusion processes: under complete control of the rate by reactions at the metal-oxide interface, or where the oxide layer maintains a constant thickness (e.g. due to its evaporation), the linear law is valid. When electron transfer occurs more slowly than ion transport, and is thus rate-determining, the logarithmic law supersedes the reciprocal-logarithmic law.

C. The oxidation rate is determined not by diffusion through the oxide lattice, but through micropores which close during oxide growth: when all pores are closed, the asymptotic growth law applies. Due to the influence of the pores during their closing, this process follows the logarithmic law.

Figure 7 shows the various oxidation laws in graphical form.

As was mentioned earlier, some oxidation laws are not applicable in the environmental temperature range. This is not true of the logarithmic and reciprocal-logarithmic laws, which are very important in this range.

At environmental temperatures, the crystal lattice defects in the oxide layer are relatively stationary. In the absence of a potential gradient, their transport is hardly detectable, and the required gradient can arise only in extremely thin oxide layers. Mott and Cabrera [6] suggest that metal ion transport through this layer occurs with the help of the so-called tunnelling effect. This transport can occur so rapidly that it is no longer the rate-determining part-process, and the transfer of metal ions into the oxide layer assumes this role. The growth rate is related to the number of cations which can overcome the necessary energy barrier:

$$\frac{dy}{dt} = c \exp \frac{\delta}{y} \tag{3-6}$$

(δ = limiting value of layer thickness, y = layer thickness, t = time, c = constant.) Assuming that $y \ll \delta$, this equation leads to the reciprocal-logarithmic law (Equation (3-2)). This derivation also assumes that the limit value of the layer thickness, δ, lies at the limit of this law. It has been confirmed experimentally for copper and aluminium at low temperatures [7, 8].

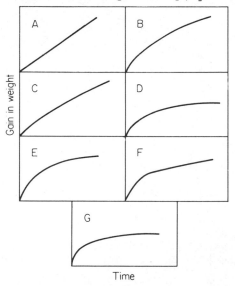

Figure 7. Schematic representations of the
laws governing oxidation rates.
A. linear; B. parabolic; C. mixed parabolic;
D. asymptotic; E. logarithmic; F. reciprocal
logarithmic; G. cubic.

The logarithmic law is valid if it is assumed that in very thin oxide layers the rate of cation transport in the metal layer is faster than electron transfer at the oxide-atmosphere phase boundary. The electron transfer rate is then rate-determining. The electrons penetrate the layer by the tunnelling effect, and the number decreases exponentially with increasing layer thickness. The rate of layer growth thus determines the number of electrons penetrating the layer:

$$\frac{dy}{dt} = C \exp(-ky) \tag{3-7}$$

18

or in integrated form (assuming that $y = 0$ at $t = 0$):

$$y = k \ln{(at + 1)} \tag{3-3}$$

Landsberg [9] has derived the same expression by assuming the rate-determining process to be oxygen adsorption. This law is valid for e.g. initial formation of oxide on zinc at 25 °C.

Weissmantel [10] has studied the kinetics of oxygen adsorption onto metals. It was shown by manometric measurements that the dependence of oxygen sorption on time follows the logarithmic law only for approximately 12 to 13 minutes. The later sorption proceeds more slowly and, like the overall

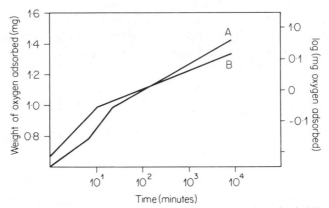

Figure 8a. Adsorption of oxygen onto iron, plotted on single (A) and double (B) logarithmic coordinates [10].

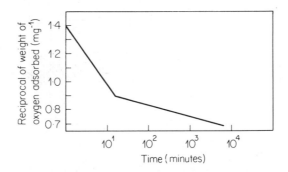

Figure 8b. Adsorption of oxygen onto iron, plotted on a reciprocal scale [10].

process at elevated temperatures, can be described by the generalized power law:

$$a = c \cdot t^{1/n} \tag{3-8}$$

The values of the constants c and n here are dependent on temperature and metal type. Weissmantel also describes the influence of water vapour on the oxygen sorption process: after an initial inhibition, the oxygen consumption increases more rapidly than in absence of water vapour. This is explained by initial separation of the oxygen from the surface by chemisorbed water, and later hydration and splitting open of the thin oxide layer. An exception to this is copper, whose oxidation at low temperatures is significantly inhibited by water vapour.

If the formation of thin oxide layers in dry atmosphere does not favour corrosion of the metal, this is of some importance in practical corrosion protection and test methods. The influence of such layers on the processes of surface protection may be estimated, and in corrosion testing the intervals

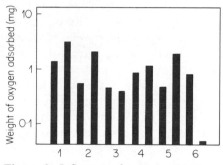

Figure 9. Influence of water vapour on oxygen adsorption (after 60 ks) onto various metals at 20 °C [10]. (Left hand column of each pair in absence of water vapour, right hand column in presence of 4·58 Torr H_2O). 1. Fe; 2. Ni; 3. Pt; 4. Cr; 5. Al; 6. Cu.

between cleaning operations (e.g. pickling) and exposure of the sample in the test medium should be specified exactly.

Many questions regarding these thin oxide layers are still unanswered; very little is known about their influence on mechanisms of adhesion of protective coatings, especially organic coatings, and on their chemical properties. Data on the influence of other species (e.g. water vapour) on the properties and the kinetics of formation, etc., of thin oxide layers are sparse.

3.2 Atmospheric corrosion as a special case of metal corrosion in an electrolyte

The empirical result that marked corrosion phenomena are found only when atmospheric humidity exceeds a certain critical threshold value leads to the conclusion that atmospheric corrosion is a type of electrochemical corrosion process. This process differs from others of its type, however, in that it occurs in a limited quantity (thin layer) of electrolyte.

Some ideas on this topic may be drawn from general knowledge of corrosion in electrolytes, and of properties of aqueous precipitates and adsorption layers in the atmosphere.

A. It is expected that the electrolyte causing corrosion has a pH value close to 7 (whereas, as was discussed in Chapter 2.2.3, dew and rain in severely polluted industrial atmospheres can have values of pH < 3, so that the stock of H^+ ions in the electrolyte layer is soon consumed by the corrosion reaction, and pH = 7 is approached again.)

B. The extremely thin electrolyte layer on a metal surface is in a continuous state of 'internal' change because of temperature and humidity fluctuations. It may therefore be assumed that in view of the high oxygen content of the air and its considerable solubility in the electrolyte layer, its transport to the metal surface to take up electrons in the cathodic part-reaction can hardly be the slowest, and hence rate-determining, step. Thus, at pH = 7, oxygen reduction is the most important cathodic part-reaction for corrosion of technical metals. In most cases the maintenance of passivity may be expected for metals which are passive under atmospheric conditions (e.g. chromium, titanium, aluminium, or Cr–Ni special steels.)

C. Since atmospheric corrosion proceeds in a very limited electrolyte layer which is only temporarily present on the surface and has an approximately neutral pH, it may be assumed that metals which are not passive under atmospheric corrosion conditions will form solid corrosion products, which will remain attached to the surface in absence of mechanical effects. This occurs because there is rapid exceeding of the solubility products of the compounds formed, supersaturation of the electrolyte, and formation of the necessary crystallization nuclei [11]. Crystallization will continue unimpeded even when the corrosion process is dormant.

D. The relatively thick and adhesive layer of corrosion products will undoubtedly affect the path of the corrosion process. The formation mechanisms of the products, their individual corrosion resistance, their content of soluble species, their structure and homogeneity, and their penetrability by corrosive atmospheric components can be particularly important in deciding the course of atmospheric corrosion. This may be true to such an extent that these layers act in concert with the formation and consolidation of the electrolyte, and its composition, in promoting the corrosion.

It is clear from the above that corrosion processes are not totally differentiable, since the concept of passivity is not exactly defined in these cases. 'Passive' in B and C implies that type of passivation which arises from thin two- and three-dimensional oxide films (as on titanium, special corrosion-resistant steels, chromium, etc.), rather than the thermodynamic definition of Pourbaix [12], which involves also corrosion with formation of thicker, more protective layers of corrosion products (e.g. on zinc, cadmium, copper, iron, etc.)

Summarizing, atmospheric corrosion may be defined thus:

Atmospheric corrosion is an electrochemical corrosion process which occurs in an alternatingly growing and decreasing layer of electrolyte, which is very thin. Depending on the type of corroding metal, atmospheric corrosion can occur in either the active or passive potential range of the metal, because of the approximately neutral character of the electrolyte. In active corrosion, it is usual to have formed a partly adherent, thick layer of corrosion products, whose properties can have significant effects on the kinetics of the process. In such a case, the metal may be described as becoming passive, in terms of the definition of passivation due to Pourbaix.

3.3 Formation and destruction of electrolyte layers

3.3.1 The formation of electrolyte layers

The first requirement for active electrochemical atmospheric corrosion is the presence of an electrolyte on the metal surface. Since, as was discussed in Chapter 2, water can be present in all its forms, all the corresponding possibilities for formation of this electrolyte must also be considered.

A. Water is present in the atmosphere in liquid form. In rain the dewpoint of the region close to the earth is exceeded, and uncondensed water vapour converts into liquid water (drops). The liquid water falling to the earth covers exposed metal surfaces with a relatively thick electrolyte layer (up to 0·2 mm thick). In rain, there is an excess of water for the corrosion process. The water layer renews itself continuously during rainfall, and the large variations undoubtedly cause intensive introduction of water-soluble components of the atmosphere (especially O_2) which aid the corrosion. On the other hand, in prolonged rain there is washing away of accumulated corrosion stimulators from the surface, and so the corrosion process may be slightly inhibited. In addition, there is washing-away of previously-formed corrosion products. These products may be sparingly soluble, so inhibiting corrosion significantly by forming a more or less continuous surface layer. The mechano-erosive action of rain is significant in this case.

The case where the surface is wetted by precipitation of very small water drops (which in mist are a type of 'aerosol') is quite different. The amount of water precipitated is far smaller here, and rarely leads to major water exchange on the surface. The huge number of small water droplets is undoubtedly saturated with atmospheric oxygen and all other species important for corrosion which are present in the atmosphere, because of the large surface area of the drops. A continuous liquid layer is visible on 'pure' metals only after prolonged exposure to mist, and this forms by coalescence of isolated drops [13]. If corrosion products already exist, a continuous water layer is likely to form during fog; this is clearly due to the microporous structure of the predominantly crystalline corrosion product. Electrolyte layers formed during exposure to fog are thinner than those in rain, but more aggressive because of the extra corrosion stimulator content and because they are not washed off again.

B. Water in the air in vapour form. The physical adsorption of water vapour at the phase boundary between pure metal and atmosphere and its conversion to liquid form (effectively an electrolyte layer) appears unlikely. Though there is only limited data on the adsorption conditions in the metal-H_2O(g) system [10], it is likely that even with multi-layer adsorption the layer thickness (approximately 2 nm) is still too low to act as a significant electrolyte. Further, in sorption of water vapour onto metals, both physical and chemical binding forces must be considered. There is evidence that water vapour is chemisorbed, in which case dissociative chemisorption can occur [14]. Such physical or chemical processes can hardly form a satisfactory electrolyte layer on the metal surface. Further, formation of such electrolyte layers by capillary condensation into small micropores and valleys on metal surfaces seems very unlikely, since capillary condensation will occur only slightly below the saturation value for atmospheric humidity [15].

It seems appropriate at this point to consider the often-raised question of the influence of microporous materials with high water vapour adsorption capacities. In the presence of such materials, e.g. soot or silica gel, on the surface, initiation of atmospheric corrosion is likely, with the water deposited on these materials even at low water vapour partial pressures acting as the effective electrolyte. However, Bartoň [16] and Buckowiecki [17] have found that this does not occur in practice. This is presumably due to the strength of the forces of water adsorption to the material.

The lowering of the dewpoint by hygroscopic water-soluble species seems much more important for the conversion of water vapour to liquid water on the metal surface. This can occur either directly at the surface of the metal or in aerosol formation in the air.

The first of these cases is more interesting from our point of view. Water-soluble salts coming into contact with humid air after being dried take up water vapour from the atmosphere. If the equilibrium pressure of the water vapour which corresponds to a solid species is exceeded, further water vapour uptake occurs to give a solution whose concentration corresponds to an equilibrium value between the water vapour partial pressure over the solution and that existing in the atmosphere (Table 4). Particularly in atmospheres which contain large quantities of such species, and in which the metal surface is more or less contaminated with them, this type of electrolyte formation is of special importance. This is especially true in coastal and industrial regions; in the former because of chloride, in the latter because of sulphate.

Aerosol-forming hygroscopic soluble solid species convert into their dissolved forms before deposition. The electrolyte (the concentrated salt solution) falls in droplets to the surface, and these coalesce to form the electrolyte required for corrosion.

Salt particles from the atmosphere are not the sole source of hygroscopic species at the metal surface. During corrosion reactions between metals and gaseous species (normally air pollutants), soluble hygroscopic products form, and these lead in turn to electrolyte formation. In this respect, sulphate (which arises chiefly as the conversion product of atmospheric sulphur dioxide) and chloride are particularly important. The new phase (the solid corrosion product) formed during the reaction affects the water participation in the reaction system. The fact that accumulations of more or less soluble salts, with anion depending on atmospheric type, are found in corrosion products is an expression of this. (This will be discussed more closely later). In industrial atmospheres there is

Table 4. Relative humidities (%) over saturated salt solutions

			Temperature °C		
Salt	10	15	20	25	30
$Pb(NO_3)_2$	99·0		98·0		96·5
KH_2PO_4	97·6	99·4	97·0	96·6	93·8
$Na_2C_4H_4O_6 . H_2O$		94·0	92·0	92·0	92·0
$Na_2CO_3 . 10H_2O$	99·0		92·0	87·0	87·0
K_2HPO_4			92·0		
$KNaC_4H_4O_6 . 4H_2O$	87·5		87·2		87·1
KNO_3		96·2		93·0	92·0
$ZnSO_4 . 7H_2O$			90·0	88·5	
K_2CrO_4			88·0		
KCl		86·7			84·6
KBr	86·0		84·0		82·0
NaCl	77·0	77·5	77·0	76·0	76·0
$NaNO_3$	77·7	77·2	77·7	75·0	73·2
$NaNO_2$			66·0		
$Ca(NO_3)_2 . 4H_2O$			55·9	51·0	49·0
$Na_2Cr_2O_7$			52·0		
KNO_2			45·0		
$MgCl_2 . 6H_2O$			34·0	33·0	31·7
$CaCl_2 . 6H_2O$	38·0		32·3	31·0	
$ZnCl_2 . H_2O$	10·0		10·0		
NaOH			5·5	6·5	4·0

thus, e.g. on steel, the formation of sulphate nests, etc. [18]. The anions raise the colloidal component of the corrosion products (especially rust). These colloids bind the water relatively loosely in liquid form below the water vapour saturation pressure, so that it is easily freed to act in the corrosion process [19].

The formation of electrolyte layers on surfaces will be discussed further in later chapters, especially with respect to the long-term course of corrosion.

3.3.2 The destruction of electrolyte layers

In contrast to the multiplicity of mechanisms of electrolyte layer formation, their disappearance can in practice be due only to evaporation or to formation of chemical reaction (corrosion) products. Water layers can evaporate from metal surfaces only at below 100% relative humidity. The evaporation process is thus dependent above all on the environmental relative humidity. A significant temperature effect is likely. Based on these considerations, Berukschtis and Klark [20] developed equation (3-9) to calculate the duration of electrolyte presence after precipitation:

$$\tau = \frac{\delta}{v} = \frac{\delta}{v_0 - (75 - H_r)\dfrac{dv}{dH_r} + (20 - t)\dfrac{dv}{dt}} \tag{3-9}$$

In this equation, τ is duration of electrolyte presence, δ is the thickness of the electrolyte layer, v is the evaporation rate at the given temperature and humidity,

v_0 is the evaporation rate at 20 °C and 75% relative humidity, and H_r and t are the actual values of relative humidity and temperature. They have determined empirical values of $v_0 = 35\,\mu m\,h^{-1}$, $dv/dH_r = -0.38$ per percent relative humidity, and $dv/dt = +0.85$ per °C, and attempted to calculate the total time of electrolyte presence using the equation. Though this method appears well-founded at first sight, it is not generally applicable; the equation assumes a general 'critical' humidity (75% relative humidity) below which no corrosion should occur, yet it is well-known that the critical humidity value depends chiefly on the chemical nature of the atmosphere and on the reactive interface of the metal system. It is further assumed that at all humidities below 100% the electrolyte is lost by evaporation, yet this is not true in the very common case where the electrolyte is a concentrated solution of hygroscopic substances, which has a definite water vapour pressure at lower humidities in equilibrium with the water vapour partial pressure of the atmosphere. It is doubtful whether the disappearance of water is exclusively due to evaporation in presence of colloidal corrosion products.

Thus, though the disruption mechanism is predominantly by evaporation, it seems that the actual relationship is much more complicated than can be expressed in a simple formula such as this.

Further considerations should also be mentioned: a corrosion process involves the existence of an electrolyte in contact with a metal. Since most corrosion processes yield solid products, which contain chiefly hydroxide, hydrated salts, and oxides, part of the water is removed from the system in a more or less solidly bound form. These products also have a relatively high sorption capacity for water, which leads to binding of a further fraction of the water. Thus, on the one hand there is partial removal by chemical processes of the very small amount of electrolyte which is effective in atmospheric corrosion (especially at relative humidities below 100%), while on the other hand there is a layer of corrosion products formed which can be regarded as a water reservoir, and which under some circumstances (e.g. by exceeding the sorption capacity as the humidity rises) can liberate liquid water. The reverse process is also possible.

The important questions involved in formation and destruction of surface electrolytes layers during atmospheric corrosion are thus far from explained, though they are of considerable importance for the long-term prediction of corrosion characteristics. (See also Chapter 3.8).

3.4   Electrochemical reactions in atmospheric corrosion

3.4.1   Atmospheric corrosion as a redox process

A close analysis of the electrochemical reaction mechanisms of atmospheric corrosion must depend mainly on general ideas on electrochemical corrosion, though the peculiarities of the process (as described in Chapter 3.2) should also be considered.

As for other cases of electrochemical metal corrosion, atmospheric corrosion should be regarded as a total process comprised of simultaneous oxidation (anodic) and reduction (cathodic) reactions. Each reaction type is tied to the other; i.e. the oxidation process (the actual corrosion) could not proceed without there being a simultaneous reduction reaction, though the individual mechanisms of the two processes may be completely independent.

In atmospheric corrosion, the electrochemical process is followed by pure chemical conversion of the corrosion products. This affects the kinetics of the primary electrochemical corrosion. More attention will be given to this in Chapters 3.7 and 3.8. A general scheme of the electrochemical processes of atmospheric corrosion is shown in Figure 10.

It is clear from Figure 10 that the corrosion proceeds as fast as is needed to balance the cathodic current in the opposite direction at the stationary

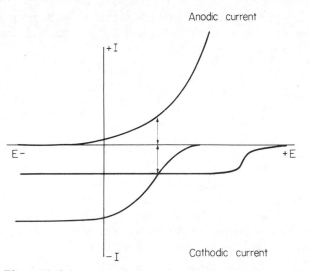

Figure 10. Schematic polarization curves for atmospheric corrosion, with two different cathodic processes.

potential of the system. Thus, different mechanisms (signified by the different polarization curves) for different reduction reactions may lead to identical corrosion rates (expressed as current densities).

It seems desirable that the factors discussed in Chapters 3.2 and 3.3, and particularly the characteristic electrolyte properties, should now be considered in examining the oxidation and reduction reactions more closely.

3.4.2 Cathodic processes in atmospheric corrosion

If, as was discussed in Chapter 3.2, it is assumed that the surface electrolyte in its extremely thin layers is neutral or, exceptionally, slightly acid, the reaction

$$2H^+ + 2e^- \rightarrow H_2 \tag{3-10}$$

need hardly be considered for atmospheric corrosion of the important metals and alloys. Even the most electronegative metal, magnesium, corrodes mainly with oxygen consumption as the reduction step under these conditions.

The reduction of atmospheric oxygen, whose solubility is high enough (about 2.5×10^{-4} molar at $20\,°C$) to provide a sufficient supply of electron acceptors to sustain the corrosion current, is thus the most important reaction in which electrons are taken up.

Cathodic oxygen reduction is basically via one of the reactions:

In acid solution $\qquad O_2 + 4H^+ + 4e^- \rightarrow 2H_2O$ (3-11)

In neutral or $\qquad O_2 + 2H_2O + 4e^- \rightarrow 4OH^-$ (3-12)
alkaline solution

The latter seems the most likely for atmospheric corrosion in an approximately neutral solution [21]. This overall reaction proceeds with transfer of 4 electrons, but various authors have found in studies of cathodic oxygen reduction in electrolytes (e.g. at amalgam electrodes [24]) that two steps are involved. (The overall reaction as a single step is unlikely because of the 4-electron transfer.) Hydrogen peroxide occurs as an intermediate:

$$O_2 + 2H_2O + 2e^- \rightarrow H_2O_2 + 2OH^- \qquad (3\text{-}13)$$

or $\qquad O_2 + 2H^+ + 2e^- \rightarrow H_2O_2$ (3-14)

and is then further reduced:

$$H_2O_2 + 2H^+ + 2e^- \rightarrow 2H_2O \qquad (3\text{-}15)$$

or $\qquad H_2O_2 + 2e^- \rightarrow 2OH^-$ (3-16)

Roich [23] showed by using the photochemistry of the hydrogen peroxide produced in this mechanism that this intermediate is undoubtedly formed in atmospheric corrosion of magnesium, zinc, nickel, and other metals. There is, in addition to this mechanism, a direct one-step oxygen reduction involving 2-electron transfer [24]. This reaction is schematically:

$$O_2 \rightarrow 2O_{ads} \qquad (3\text{-}17)$$

$$O_{ads} + 2H^+ + 2e^- \rightarrow H_2O \qquad (3\text{-}18)$$

or $\qquad O_2 + 2e^- \rightarrow O^{2-} + O_{ads}$ (3-19)

which is alleged to proceed catalytically [25]. Since, as will be shown later, corrosion products exert an important catalytic effect on the mechanism, these reactions must be considered in each case. Proof that both reaction schemes occur simultaneously has been provided by Gerasimov and Rosenfeld [24], whose calculations show electron consumption values (from limiting currents at a rotating electrode) of between 2 and 4.

Some still-open questions on cathodic reduction will not be discussed here. The irreversibility of the process rules out comparison between oxygen consumption in atmospheric corrosion and oxygen consumption at the oxygen electrode according to different theories. The same applies to the proposed mechanisms involving intermediate radicals and ions such as HO_2 or HO_2^- [21]. It should be noted that the oxygen reduction proceeds at low current densities (10^{-1} to $1\ mA\ cm^{-2}$) according to the Tafel equation:

$$\eta = a + b \log i \qquad (3\text{-}20)$$

(where η = overpotential, i = current density, and a and b are constants). Tomashov's measurements of overpotentials on different metals show a considerable dependence of the constant a on the character of the metal (Table 5). The constant b should be calculable from the expression $2RT/F = 116\,\text{mV}$, and be independent of material. The highest measured values may be attributed to the presence of oxide (hydroxide) layers, which either inhibit the reaction

Table 5. Values of overpotential for electrochemical reduction of oxygen [21]

Metal	Overpotential (V)	
	at $i = 5\,\text{A m}^{-2}$	at $i = 10\,\text{A m}^{-2}$
Pt	0·65	0·70
Au	0·77	0·85
Ag	0·87	0·97
Cu	0·99	1·05
Fe	1·00	1·07
Ni	1·04	1·09
C (graphite)	0·83	1·17
Cr	1·15	1·20
Sn	1·17	1·21
Co	1·15	1·25
Fe_2O_3	1·11	1·26
Cd	1·28	
Pb	1·39	1·44
Ta	1·38	1·50
Hg	0·80	1·62
Mg	2·51	2·55

directly (e.g. by their Ohmic resistance) or partly screen the surface so that the actual current density is higher than it appears to be. The varying overpotentials for cathodic oxygen reduction on oxide-covered metals undoubtedly play an important role in atmospheric corrosion. As long as the oxygen penetrates

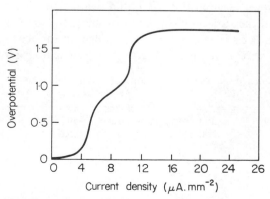
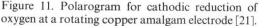

Figure 11. Polarogram for cathodic reduction of oxygen at a rotating copper amalgam electrode [21].

the electrolyte layer by diffusion, a limiting current region of the cathodic polarization curve should be expected, though the actual diffusion process is rapid (Figure 11). Using the diffusion coefficient and Fick's first law, the following expression can be deduced for the limiting current density, i_D, (assuming that $c_{O_2} = 0$ at the metal surface):

$$i_D = \frac{DnFc}{\delta} \tag{3-21}$$

(where D = diffusion coefficient, F = Faraday constant, c = oxygen concentration in the electrolyte, δ = diffusion-limited layer thickness and n = the number of electrons passed through the boundary surface). The diffusion coefficient may be reasonably defined by

$$D = \frac{RT}{N} \cdot \frac{1}{6\pi\eta r} \tag{3-22}$$

(where R = gas constant, T = absolute temperature, N = Avogadro's number, η = electrolyte viscosity, r = molecular radius of oxygen). Substituting for D in Equation (3-21),

$$i_D = \frac{RTnFc}{6\delta N\pi\eta r} \tag{3-23}$$

Rosenfeld and Zhigalova [26] have performed an experimental test of this equation, which expresses the dependence of the limiting current on the diffusion layer thickness. Their measurements show (Table 6) that diffusion transport is likely through electrolyte layers thinner than approximately 30 μm under strictly isothermal conditions.

In this range, the electrolyte layer thickness and the effective diffusion-limited layer are approximately comparable, but in thicker layers and especially under non-constant temperatures there is a strong convective effect, which often leads to higher values of limiting current density than those calculated. Convection arises because of local temperature differences, which may be due to e.g. thermostatic evaporation of liquid water (the higher the humidity gradient between the surface and the atmosphere, the greater this effect), and can be explained as capillary convection. Changes in surface tension because of local temperature or concentration gradients also cause convection. Levich

Table 6. Dependence of oxygen reduction rate on the thickness of the electrolyte layer (0·1 M NaCl).

Layer thickness μm	Limiting current density, A m^{-2} calculated	measured	Diffusion layer thickness, μm
640	0·293	1·25	152·0
320	0·596	1·55	122·3
160	1·191	1·72	111·0
70	2·730	3·20	59·6
30	6·355	6·20	30·8

(from [21]) calculates a transport rate of 0.1 cm s^{-1} for a layer thickness of $300 \, \mu m$ and a temperature decrease of 0.1 deg cm^{-1}. Measurements indicate that the limiting current density of oxygen reduction in thin electrolyte layers may exceed $600 \, \mu\text{A cm}^{-2}$ (for a layer thickness of $30 \, \mu m$) [26]. This would correspond to a possible corrosion rate on iron of $6.2 \text{ g m}^{-2} \text{ h}^{-1}$, but this exceeds that found experimentally in long-term corrosion investigations by several orders of magnitude. (Bartoň and Bartoňová [27] quote values of 0.02 to approximately 0.1 (and exceptionally 0.5) $\text{g m}^{-2} \text{ h}^{-1}$). Everything thus seems to indicate that atmospheric corrosion is in no way controlled by the cathodic oxygen reduction process.

Electrochemical reduction of oxygen is undoubtedly the most important cathodic process in atmospheric corrosion. However, since the metal-atmosphere system is a complex redox system, in which the corrosion products being formed as a new phase should not be neglected, the possibility of their participation in the cathodic reactions should be considered. This applies in particular to those metals whose corrosion products can arise in several discrete steps. For example, it can be shown that in the $\text{Fe}-\text{Fe}^{2+}-\text{Fe}^{3+}$ system, corrosion can take place in absence of oxygen, in which case the Fe^{3+} ions act as electron acceptors. The reaction

$$2\text{Fe}^{3+} + \text{Fe} \rightarrow 3\text{Fe}^{2+} \tag{3-24}$$

is thermodynamically feasible, and has been demonstrated [28]. However, it should be regarded as an important process in atmospheric corrosion only when, for example, the surface electrolyte layer is so thick that the oxygen supply to the surface is very markedly inhibited (e.g. when the surface corrodes alternately underwater and in the atmosphere above it). Results suggest that atmospheric corrosion should be regarded as a multi-step redox system. In the $\text{M}-\text{MO}-\text{MO}_x-\text{O}_2$ system, the different interfaces have different oxygen and electron supplies. Atmospheric oxygen can be converted at the corrosion product-air phase boundary into oxide ions which are primarily chemisorbed, and then either oxidize lower-valent metal compounds after transport into the interior of the corrosion product or react with mobile metal ions at the surface. (The former mechanism appears more likely on iron.)

This applies not merely to metals whose corrosion products have several valencies. It can be shown that on those metals whose corrosion products have only one valency (e.g. zinc), similar theories are valid. At the $\text{Zn}-\text{ZnO}$ (or Zn(OH)_2, as applicable) interface, the excess of zinc ions is wholly absorbed into the oxide and decreases in the direction of the outer phase boundary (oxide-atmosphere). This was illustrated by measurement of the photo-effect in these semi-conductor systems [29].

All this, however, does not govern the atmospheric corrosion kinetics; the permanent excess of oxygen dissolved in the electrolyte is sufficient, especially on already-existing corrosion layers, to ensure maintenance of stationary redox conditions, so that the cathodic step has an invariant rate (from a kinetic viewpoint), and is definitely not the slowest, and hence cannot be the rate-determining, step.

3.4.3 Anodic processes in atmospheric corrosion

The anodic dissolution of metals is a process which is so far not completely explained. Though it has been known for a long time that the actual transfer reaction (ionization) is the slowest and therefore rate-determining part-process in the simplified reaction:

$$Me \rightarrow Me^{n+} + ne^-$$ (3-25)

this reaction cannot be divided unequivocally into a multi-step mechanism. The details of the theories of Bockris, Heusler, Schwabe, and Florianovich, Kolotyrkin et al [30 to 34] will not be discussed here, though it should be mentioned that their theories, which deal with the reaction components H^+, OH^-, and other ions, are very important for the interpretation of atmospheric corrosion kinetics. (See Chapters 3.5 and 3.8). For our study, it suffices to know that the overall transfer reaction mentioned above is the slowest and that, using generally-valid laws of chemical kinetics, the rate of this potential-dependent reaction can be described by the equation:

$$i = k \exp \frac{\alpha n F \varphi}{RT}$$ (3-26)

where i = current density, k = rate constant, α = coefficient, n = number of electrons transferred, φ = electrode overpotential, F = Faraday constant, R = thermodynamic gas constant, and T = absolute temperature. By taking logarithms and gathering all constants together, the Tafel equation (Equation (3-20)) is obtained:

$$\eta = a + b \log i$$ (3-20)

where η = overpotential and a and b are constants. This expresses the potential-current density relationship in anodic dissolution of most metals. Constant a is chiefly a function of the metal; b corresponds to the expression RT/nF, and should be independent of the electrode nature. Deviations from these theories may be caused by amalgamation of individual steps of the total anodic reaction, as has been investigated and described by Bockris, Heusler, etc. [30 to 34].

There is also the case where anodic polarization causes passivation. Without going more closely into the different theories of passivity (for which see e.g. Hoar, Schwabe, Uhlig, etc. [35]), it should be stated here that for atmospheric corrosion with formation of (visible) layers of corrosion products, the thermodynamic definition of passivity, as used by Pourbaix for the calculation of pH-potential diagrams, appears highly useful [12]. As was mentioned earlier, conditions are favorable for formation of solid, semi-adherent corrosion products on non-passive metals during atmospheric corrosion; using the Pourbaix definition, atmospheric corrosion must thus mostly occur in the passive region. (Obviously, this can only be true in the absence of passivity-destroying anions from the electrolyte (see Chapter 3.2.))

Metals which are very easily passivated, such as stainless, high-alloy steels, titanium, chromium, aluminium, etc., may exceptionally exhibit localized

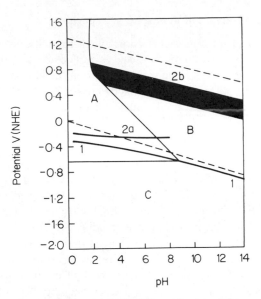

Figure 12. Potential-pH diagram for iron (after Pourbaix). 1. Oxygen-free solution; 2a. Clean iron in oxygen-saturated solution; 2b. Rusted iron in oxygen-saturated solution; A. Corrosion region; B. Passive region; C. Immune region.

depassivation, due to effects of atmospheric corrosion, leading mainly to pitting corrosion.

These ideas are basically confirmed by polarization measurements in thin electrolyte layers and by Rosenfeld's thermodynamic calculations [21].

It can be shown easily for copper, zinc, iron, and nickel that in addition to the simple transfer reaction:

$$Me \rightarrow Me^{n+} + ne^- \tag{3-25}$$

the direct electrochemical formation of solid corrosion products by the scheme (written for a di-valent metal):

$$Me + 2OH^- \rightarrow MeO + H_2O + 2e^- \tag{3-27}$$

$$Me + 2OH^- \rightarrow Me(OH)_2 + 2e^- \tag{3-28}$$

is also important. Solid secondary corrosion products may also screen the metal surface, increasing the passivity from a thermodynamic point of view:

$$Me^{2+} + 2OH^- \rightarrow Me(OH)_2 \tag{3-29}$$

The influences of the anodic processes on kinetics of atmospheric corrosion are examined more closely in Chapters 3.7 and 3.8.

3.5 Mechanisms of action of atmospheric impurities

The acceleration of atmospheric corrosion by air impurities has been recognized for a long time. It is particularly noticeable in coastal and industrial regions, since, in the first case, chloride from the sea and, in the second, gaseous or solid emissions play an important role. As was discussed in Chapter 2, the impurities in industrial regions are manifold, but the most important effect is from the extremely widespread pollutant, sulphur dioxide.

3.5.1 Chloride

The corrosion promotion by chloride ions in coastal areas is doubtless the basis for the original, and still very popular, corrosion test using exposure to an artificial salt mist. Though the stimulating action of relatively small amounts of chloride has long been known, the mechanism of the effect is still far from clear. Besides the no doubt correct assumption that the chloride in hygroscopic species e.g. NaCl, $CaCl_2$ or $MgCl_2$, helps to promote the electrochemical course of atmospheric corrosion by favouring electrolyte formation at relatively low values of relative humidity, direct participation of chloride in the anodic mechanism is also likely. In special cases, it is known that chloride (and similarly other halides) produces pitting or stress corrosion cracking of easily passivated metals and alloys. In these cases, there is simultaneous primary adsorption of passivating OH^- ions and activating Cl^- ions, which obeys a Freundlich or Langmuir adsorption isotherm. The degree of disruption of local passivation is closely related to the inhomogeneity of the passive layer on the surface, which is in turn influenced by lattice disruption at the metal surface.

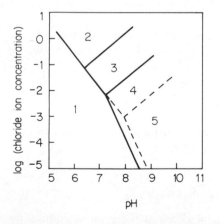

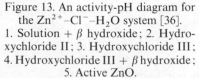

Figure 13. An activity-pH diagram for the $Zn^{2+}-Cl^--H_2O$ system [36].
1. Solution + β hydroxide; 2. Hydroxychloride II; 3. Hydroxychloride III;
4. Hydroxychloride III + β hydroxide; 5. Active ZnO.

Since atmospheric corrosion of most metals occurs in the (thermodynamic) passive region (though the passive state here is very labile), chloride (and other types of ion) might be expected to have a similar mechanism of action on them all. As chloride content in the surface electrolyte is increased, the corrosion rate should increase, because of surface depassivation. In the case of atmospheric corrosion, the rate of corrosion should thus be a function of chloride ion activity.

The activity of the chloride ions demands special mention, since it is strongly affected by the reaction products. Feitknecht [36] in particular has shown that the corrosion products can bind large amounts of chloride relatively tightly (e.g. as basic salts). Conclusions on the kinetic effects of chloride may be drawn from the stoichiometric composition and solubility products of these corrosion products. Thus it can be deduced, for example, that iron, which forms no stable basic chloride, is affected by action of chloride far more strongly than metals such as zinc, cadmium, or copper, whose basic salts of different compositions are only slightly soluble.

Finally, it can be shown that in the action of chloride on iron, (e.g. in coastal areas), chloride accumulations (nests) are formed which contain $FeCl_2$. These occur uniquely on iron [36].

3.5.2 Sulphur dioxide

It is not surprising that clarification of the mechanism of action of this widespread industrial air pollutant has been sought for such a long time, since it has a pronounced corrosive effect. Better knowledge of its conversion could lead to more effective corrosion protection measures. Schikorr [38] has written a good review of the published literature, which summarizes especially the empirical viewpoints on this theme. Basically, there are two theories which seek to explain the mechanism of the accelerating effect of sulphur dioxide: one suggests an effect on the cathodic process [39], while the other, more widespread, theory sees it having a decisive influence on the anodic process and the associated formation of solid corrosion products, which is particularly important in the long-term course of atmospheric corrosion. There is also a hypothesis of a secondary, purely chemical, action of sulphur dioxide (or its conversion products).

The theory of acceleration by sulphur dioxide of the cathodic reaction, and hence of the whole process, was proposed by Rosenfeld [21]. On the basis of polarization measurements, he described the accelerating effect of the sulphur dioxide as increasing the cathodic reaction rate by the reaction:

$$2SO_2 + 2e^- \rightarrow S_2O_4^{2-} \tag{3-30}$$

Though cathodic reduction of sulphur dioxide appears thermodynamically possible, it is not important in atmospheric corrosion. Other authors have shown that this reaction can occur only at high partial pressures of the gas in air, at concentrations of 0.5% or more [40]; i.e. at concentrations many times greater than those occurring naturally.

The action of sulphur dioxide on the anodic reactions is closely connected with its conversion mechanism. This idea has been discussed by a series of authors [39 to 44].

Older hypotheses [42] which have been scrutinized recently by McLeod and Rogers [43] clearly do not apply to atmospheric corrosion conditions, which are characterized by a permanent excess of dissolved oxygen, though they are still advocated by Ross and Callaghan [45]. These authors include a primary disproportionation reaction via

$$2SO_2 + 4e^- \rightarrow S^{2-} + SO_4^{2-} \tag{3-31}$$

and thus, e.g.:

$$2Fe + 2SO_2 \rightarrow FeS + FeSO_4 \tag{3-32}$$

The ferrous sulphide is then oxidized to sulphate, according to this theory:

$$S^{2-} + 2O_2 \rightarrow SO_4^{2-} \tag{3-33}$$

e.g. $\quad FeS + 2O_2 \rightarrow FeSO_4 \tag{3-34}$

The reaction could lead to different intermediates, such as thiosulphate, polythionates, sulphur, etc. Sulphide can certainly be demonstrated in corrosion products under certain conditions [42, 43] e.g. in laboratory investigations, using sulphur dioxide concentrations perhaps an order of magnitude greater than occur in actual industrial atmospheres. It is difficult to rule out a direct effect of hydrogen sulphide (e.g. near to oil refineries, coke works, etc.)

Recently, several authors have considered the mechanism of the sulphur dioxide conversion reaction [44, 86–91]. The primary step must undoubtedly be a sorption process. It can be shown that the SO_2 sorption occurs differently on different metals [86–88]; on iron, sorption is typically non-uniform, and limited to discrete sites on the surface, while there is more or less even coverage of copper, zinc, aluminium and silver surfaces. The sorption rate is dependent on the atmospheric humidity and on time, and increases with increasing H_2O partial pressure. This humidity dependence appears to be related to the concept of 'critical humidity'; only if the water vapour pressure is greater than 60 to 70% of the saturation pressure is the sorption rate measurable, and it then rises steeply.

The formation of corrosion products at the surface introduces the time dependence of the SO_2 uptake. The rust on iron, which is at first localized but later spreads across the surface, promotes enhanced adsorption rates. Thus, at high humidity ($>96\%$) and a high degree of sulphur dioxide contamination of the air ($>10^{-3}\%$), the uptake is such that all SO_2 molecules coming into contact with the surface are taken up. Analyses of the corrosion product (the rust) and the surface electrolyte suggest that there is rapid oxidation by the reaction

$$\left.\begin{array}{l} SO_2 + O_2 + 2e^- \rightarrow SO_4^{2-} \\[2mm] \text{or} \quad 2HSO_3^- + 1\tfrac{1}{2}O_2 + 2e^- \rightarrow 2SO_4^{2-} + H_2O \end{array}\right\} \tag{3-35}$$

with the electrons coming from the reaction

$$2Fe^{2+} \rightarrow 2Fe^{3+} + 2e^- \qquad (3\text{-}36)$$

or, assuming that the rust layer conducts electrons, from the metal dissolution reaction (as a special case of reaction 3-25):

$$Fe \rightarrow Fe^{2+} + 2e^-$$

In contrast to iron, corrosion products on zinc, copper, and especially on aluminium lessen the rate of SO_2 uptake. As Schikorr had already shown earlier [92], there is pronounced 'SO_2 rejection' by aluminium.

It should be noted here that the conversion of SO_2 to sulphate should be expected to take place particularly at the corrosion product-atmosphere phase boundary, following primary adsorption of SO_2 and O_2. Electron transport to this phase boundary is assumed to occur by supply from the primary step of the anodic process; atmospheric corrosion products are in most cases electron-conducting [46]. An exact definition of the corrosion product-atmosphere phase boundary is difficult, however, especially at coarse-grained products such as rust. In such cases it is likely that this boundary lies inside the surface system, which facilitates electron transport further. Knotková also showed that the pH value of the surface electrolyte in atmospheric corrosion lies close to 7. Even at extremely high SO_2 concentrations (25 ppm), no pH value below 3·5 was found on pure metals, whereas all electrolytes on pre-corroded metals which were investigated had pH values greater than 5·8. At low SO_2 concentrations (1 ppm), this result was still clearcut, as expected. Reaction 3-35, which takes up electrons from the system, can be interpreted as the cathodic reduction process, and, as was suggested by Rosenfeld, can be regarded as the reason for acceleration of the cathodic process. However, as has been emphasized repeatedly, these partial processes are hardly the slowest, and thus rate-determining, steps in atmospheric corrosion, especially when corrosion product layers are already present.

The accelerating effect of sulphur dioxide is thus not due to the amount of gas dissolved in the electrolyte or the acidification of the electrolyte, but involves the sulphate ions formed by conversion of sulphur dioxide. Heusler and Florianovich et al. have devised reaction mechanisms for this conversion which seem well justified for iron, nickel and cobalt, in particular [31, 34]. Floriano-vich, Sokolova and Kolotyrkin showed that on iron the anodic dissolution current at constant electrode potential is a function of not just the OH^- activity (as supposed by Heusler), but rather of the summed activities of OH^- and SO_4^{2-} ions. The reaction scheme was described as follows:

$$Fe + H_2O \rightarrow Fe(OH^-)_{ads} + H^+ \qquad (3\text{-}37)$$

$$Fe(OH^-)_{ads} \rightarrow Fe(OH)_{ads} + e^- \qquad (3\text{-}38)$$

$$Fe(OH)_{ads} + SO_4^{2-} \rightarrow FeSO_4 + OH^- + e^- \qquad (3\text{-}39)$$

$FeSO_4$ is known to be a rust component which is always formed in industrial atmospheres, and is found particularly in layers at the metal surface (e.g. in the form of sulphate nests).

According to Schikorr [47] and Bartoň and Beránek [42], oxidative hydrolysis of $FeSO_4$ leads to formation of hydroxides as the chief component of rust:

$$FeSO_4 + H_2O \rightarrow FeOOH + SO_4^{2-} + 3H^+ + e^- \qquad (3\text{-}40)$$

Thus, the stimulating sulphate ions are reliberated, so that their activity is independent, within certain limits, of the varying atmospheric SO_2 concentration.

The acidification of the electrolyte which should result from this reaction is probably counteracted by the buffer capacity of the oxide and hydroxide corrosion products already existing on the surface, so that it may be regarded as insignificant; this also is in line with the results of Knotková quoted earlier.

A possible method for enlargement of sulphate nests is the 'overgrowth' of small nests into larger ones by a diffusion mechanism. It can be shown that in spite of the sulphate enrichment at the iron-rust interface, average sulphate concentration in the rust remains constant. There are similar profiles for the dependence of sulphate concentration and apparent rust density on distance from the interface [48]. Figures 14 to 16 are the results of leaching-out experiments on sulphate nests, and show that sulphate ion transport occurs most probably by membrane diffusion. The sulphate nest acts as a store of SO_4^{2-} ions. Further corrosion without further addition of sulphate leads to just the same leaching-out characteristics of the system as before (Figure 17). It is likely that the sulphate nest spreads during the corrosion period, on the one hand by

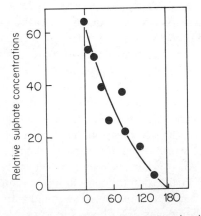

Distance from iron-rust interface (μm)

Figure 14. Distribution of sulphate in rust (determined radiochemically). (O = iron-rust interface, 180 μm = rust-atmosphere interface).

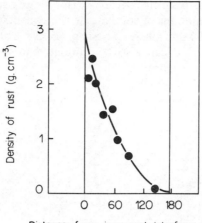

Figure 15. Density profile of rust on an iron
surface.

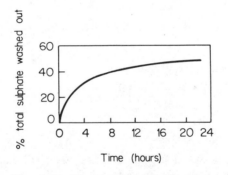

Figure 16. Kinetics of leaching of sulphate from rust on iron (measured radiochemically).

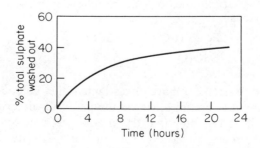

Figure 17. As for Figure 16, after further corrosion without addition to stimulator.

38

diffusion of sulphate ions from the already-existing nests, and on the other by introduction of a fresh amount of sulphate as the conversion product of SO_2 just taken up from the atmosphere. These processes occur not only under atmospheric conditions, but can also be simulated in the laboratory using radiochemical techniques (Figure 18).

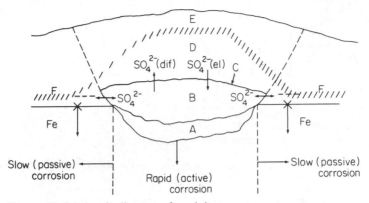

Figure 18. Schematic diagram of a sulphate nest.

A. $FeSO_4$ (solid) reactions:

$$Fe + SO_4^{2-} \rightarrow FeSO_4 + 2e \tag{1}$$

$$2Fe^{3+} + 2e \rightarrow 2Fe^{2+} \tag{2}$$

B. Solution reactions of $FeSO_4$:

$$xFeSO_4 + yH_2O + zO_2 \rightarrow xFe(OH)_2, Fe(OH)_3,$$
$$FeOOH, Fe_3O_4 + xH_2SO_4 \tag{3}$$

$$Fe^{2+} + 2H^+ + \tfrac{1}{2}O_2 + e \rightarrow Fe^{3+} + H_2O \tag{4}$$

$$SO_2 + O_2 + 2e \rightarrow SO_4^{2-} \tag{5}$$

C. Reactions 4 and 5 occur at the $Fe(OH)_3/FeOOH$ membrane.
Electrolytic transport of sulphate inwards.
Diffusion transport of sulphate outwards.

D, E. Reactions 4 and 5 occur in these regions due to presence of $FeSO_4$ and $\alpha, \gamma FeOOH$ (and Fe_3O_4?). The sulphate is washed out during period of rain.

F. There is rust formation by reaction 6 in areas where there is no sulphate present:

$$2Fe + 2H_2O + O_2 \rightarrow 2Fe(OH)_2 \tag{6}$$

$$2Fe(OH)_2 + \tfrac{1}{2}O_2 \rightarrow 2FeOOH + H_2O$$

Bartoň and Bartoňová have investigated the kinetics of atmospheric corrosion of steel, interpreting them by the processes described by Equations (3-37) to (3-40) [41]. Assuming that the anodic part-reaction is the rate-determining step, the process is second order in water activity (relative humidity) and first order in sulphate ion activity (SO_2 concentration in the atmosphere). This analysis is based on the following:

In the reaction system described by Equations (3-37) to (3-40), the processes (3-39) and (3-40) are slow, and hence kinetically significant. The surface concentration of the $Fe(OH^-)_{ads}$ complex (in Equation (3-37)) and the course of Equation (3-40) are thus dependent on the water activity in the system. One should therefore expect a second-order dependence on water activity. This agrees with measurements of the humidity dependence of the rusting process (Figure 19).

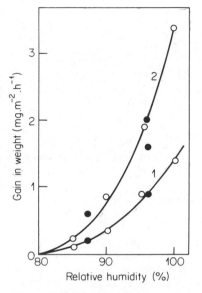

Figure 19. The dependence of rusting rate on humidity. 1. Low stimulator concentration in the rust; 2. High stimulator concentration in the rust.
○ calculated values; ● measured values.

The linear dependence of the reaction rate on sulphate activity is derived similarly. In investigations of the rusting kinetics in the presence of varying initial surface sulphate concentrations, in the form of $FeSO_4$, it was shown that the corrosion process is influenced by the activity of this stimulator. Its gradual 'dilution' in the growing rust layer leads to falling corrosion rates, so that the stationary final value is independent of the initial sulphate concentration. Establishment of the stationary corrosion rate takes longer for high initial values than for low ones. When the sulphate is applied to rusted iron (rust without sulphate content), the final rates found are lower (Figures 20 and 21).

It can be deduced from these results that the reaction is essentially first-order in surface sulphate concentration. This holds good, however, only above a certain limiting concentration required for these ions to have an effect on the reaction mechanism. As SO_4^{2-} activity in the growing rust layer decreases, this

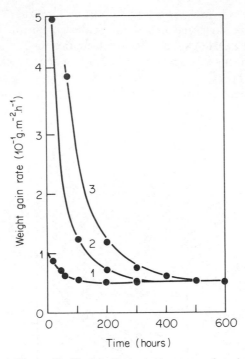

Figure 20. Changes in corrosion rates of steel with time, using different surface sulphate concentrations (and at 25 °C and 96% relative humidity). 1. 0·19; 2. 0·95; 3. 7·59 g SO_4^{2-} m^{-2}.

value is no longer exceeded, and the stationary corrosion rate, independent of the initial sulphate concentration, is very close to the rusting rate in absence of sulphate. The values found in these experiments (0·015 to 0·025 g m^{-2} h^{-1}) correspond to those measured under pure atmospheric conditions (approximately 0·025 g m^{-2} h^{-1}) (see Chapter 3.8). It can be deduced from this that the corrosion rate of iron is the sum of two concurrent reaction rates; there is a rapid reaction with participation of sulphate ions formed from atmospheric SO_2 and a slow one without this influence. If a critical sulphate concentration in the boundary layer is exceeded (this value depending on SO_2 concentration and opposition to its action from the rust layer), the contributions from the two reactions depend on the sulphate activity.

This result assumes a surface segregation of the two reactions. This is in total agreement with the phenomenon of so-called 'sulphate nests', as have been described and interpreted by Schwarz [18]. For their formation, it must be assumed that the primary step is a surface-differential adsorption of sulphate ions, for example at discontinuities in the inherent oxide layer. In the sense of an adsorption isotherm, therefore, the number of such sites (and hence raised SO_4^{2-} concentration) and the total sulphate ion activity on the surface should be related. The 'rapid' reaction mechanism occurs at negative potentials, setting up anodic areas, and the resulting electric field causes further concentration of

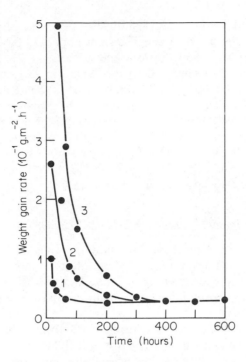

Figure 21. As for Figure 20, using pre-rusted
steel specimens which have a low stimulator
concentration in the existing rust.

the sulphate ions. Periods (e.g. in summer) when there is little addition of SO_2 lead to redistribution of sulphate through the rust layer again.

Thus, it may be concluded that, for iron, corrosion acceleration by SO_2 or sulphate ions is related above all to the primary anodic step of the corrosion.

Rather different assumptions must be applied to non-ferrous metals, as has been shown by Feitknecht, in particular [36]. In contrast to iron, predominantly oxides and hydroxides are now formed at the metal-corrosion product interface, even in heavily polluted atmospheres. The assumption that sulphate ions, rather than the SO_2 which is converted into them, are the stimulators still applies to these metals [44, 48]. The appropriate metal sulphate (normal or basic) is found on zinc, copper, etc., only at the corrosion product-atmosphere interface, and arises as a secondary product:

$$Me + H_2O + \tfrac{1}{2}O_2 \rightarrow Me(OH)_2 \tag{3-41}$$

$$Me(OH)_2 + SO_4^{2-} \rightarrow MeSO_4 + 2OH^- \tag{3-42}$$

These equations also apply, of course with different stoichiometries, for the formation of basic sulphates. These are formed in industrial atmospheres on copper ($CuSO_4 . 3Cu(OH)_2$), and also form on nickel and other metals [50 to 52].

Since the sulphates of all important metals are water-soluble, large quantities of water will wash them from the surface, and the initial thickness of the corrosion product layer must be re-established by primary hydroxide (or oxide) formation, as was discussed earlier. The same applies to basic sulphates, which are converted to more soluble forms at the surface.

These ideas have been particularly thoroughly investigated in the corrosion of zinc, cadmium and copper [42]. The higher the SO_2 content of the atmosphere, the lower the hydroxide and oxide content of the corrosion products and the higher the sulphate content. Under extreme conditions, e.g. on zinc in a railway tunnel (with steam trains), only $ZnSO_4 . 7H_2O$ is found as corrosion product. Similarly, mainly $CuSO_4 . 5H_2O$ is formed on copper in artificial atmospheres with $>0.5\%$ SO_2 [42]. Biestek demonstrated $CdSO_4$ on cadmium in an industrial atmosphere using an x-ray method [51].

3.5.3 Hydrogen sulphide, hydrogen chloride and chlorine

All reactions promoted by these air pollutants must be regarded as special cases of atmospheric corrosion, since their effects, though very intensive, are related invariably to locally-occurring industrial emissions. Little effort has so far been devoted to examining the mechanisms of their corrosion-promoting reactions.

Hydrogen sulphide is extremely reactive, and capable of reaction with most technical metals. The energy of formation of sulphides is high, and it is well-known that even relatively stable metals such as copper or silver will react with very small amounts of hydrogen sulphide to form sulphide tarnish layers. This is particularly significant in those cases where these sulphide layers impair the electrical properties of the material.

Corrosion with primary formation of sulphides can, under certain conditions, occur very rapidly. Thus, a copper corrosion rate of up to $140\,\mu m\,y^{-1}$ has been measured near a hydrogen sulphide source, with mainly CuS as the corrosion product [42]. Other metals with a high affinity for sulphide ions also corrode very rapidly in atmospheres containing hydrogen sulphide, mostly with the corresponding sulphide as the primary product which is then converted to other compounds by secondary reactions. In the rust formed on the iron samples exposed to an H_2S-rich atmosphere, as were discussed earlier (rusting rate $200\,\mu m\,y^{-1}$), 3 to 5 % of each of S(-II), S(O) and S(VI) were found.

S(-II) can also be detected in corrosion products on lead, cadmium and zinc under such conditions, but always in addition to sulphate. Rosenfeld [21] suggests that S^{2-} ions accelerate the anodic reaction, in particular.

Feitknecht and Bartoň, in particular, have studied hydrogen chloride vapour as an atmospheric corrosion stimulant [36, 53]. It should be noted, however, that this gas has been studied only as a model pollutant, in examining the difficulties involved in the complicated conversion reactions of sulphur dioxide. The experimental results found by these authors may be summarized as follows:

In the presence of atmospheric hydrogen chloride (in concentrations of 10^{-4} to 10^{-2} volume percent), the gas is first adsorbed, and reacts only if a critical humidity is exceeded, to yield the corresponding metal chloride. The critical humidity required is approximately the vapour pressure over a saturated solution of the particular chloride. Further reactions yield the basic chlorides and hydroxide (oxide) compounds. Their respective ratios can be represented graphically, using surface electrolyte pH and chloride activity as coordinates. In this way, Feitknecht and Grauer have calculated stability regions for different corrosion products on copper and zinc (e.g. [54]). It should be noted here, however, that in the metal-corrosion product-atmosphere system no uniform electrolyte properties may be expected; rather, they will change depending on the properties of the corrosion product layer formed, particularly the capacity for binding Cl^- ions by forming stable, sparingly soluble compounds. In this respect iron is a special case, since it cannot bind chloride ions into sparingly soluble compounds [53]:

$$FeCl_2 + H_2O + \tfrac{1}{2}O_2 \rightarrow FeOOH + 2HCl \tag{3-43}$$

$$Fe + 2HCl + \tfrac{1}{2}O_2 \rightarrow FeCl_2 + H_2O \tag{3-44}$$

On the other hand, corrosion is slower of those metals which form stable, slightly soluble basic chlorides, which can be represented e.g. for zinc by

$$(5 + a)Zn + 2HCl + (3 + a)H_2O + (\tfrac{1}{2} + \tfrac{1}{2}a)O_2$$

$$\rightarrow ZnCl_2 \cdot 4Zn(OH)_2 + aZn(OH)_2 \tag{3-45}$$

$$aZn(OH)_2 + 2aHCl \rightarrow aZnCl_2 + 2aH_2O \tag{3-46}$$

In this, it must be assumed that the HCl vapour participates at the corrosion product-atmosphere interface, so that there are different pH and chloride concentration here from those at the metal-product interface. It is thus understandable why the chloride concentration decreases with distance from the outer

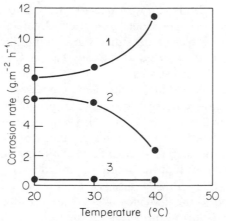

Figure 22. Effects of temperature changes on stationary corrosion rates of metals in an atmosphere containing 16 mg HCl(g) m^{-3}, at 80% relative humidity. 1. Iron; 2, Zinc; 3. Copper.

interface, and why the visibly green patina on copper is a thin layer on a compact Cu_2O-CuO foundation. Zinc behaves similarly. The corrosion products on these metals prevent the transport of corrosion stimulators to the metal surface. It is easily understood why the ratio of partial pressures $p_{HCl(g)}/p_{H_2O(g)}$ is of importance for the corrosion [53]. The higher the ratio, the greater the extent of conversion of hydroxide compounds into soluble, basic to normal chlorides. At low ratios, the formation rate of hydroxides is greater, since the chloride effect is smaller, so that the chemical reaction must proceed via an adsorption process which is necessarily affected by this ratio. There is therefore no temperature dependence of rates of copper and zinc corrosion at constant relative humidity and p_{HCl}.

As was mentioned earlier, practical use of these theories of HCl-polluted atmospheres is not common, but they are useful in taking a general view of the effects of atmospheric impurities, the more so in that HCl often acts in conjunction with the most technically-important stimulator, sulphur dioxide.

Chlorine is rarely responsible for intensive corrosion promotion in the atmosphere. Chlorine occurs in significant concentrations only around certain chemical works (see Chapter 2). In examining its mode of action, it should be tested whether it is influencing the anodic or cathodic part-processes. The solubility of chlorine in water is high, and it reacts with water by

$$Cl_2 + H_2O \rightarrow HCl + HClO \tag{3-47}$$

The ClO^- anion can promote the cathodic reaction on account of its effect on the redox potential and its ease of reduction:

$$ClO^- + 2H^+ + 2e^- \rightarrow H_2O + Cl^- \tag{3-48}$$

The activating chloride ions can then influence the anodic reaction in particular. Rosenfeld [21] says both effects must be considered, since there is an effect from surface electrolyte acidification, especially in the initial stages. If there are significant layers of corrosion products present, which can regulate the electrolyte pH and the stationary corrosion potential, the action of the chlorine should be found mainly in the anodic process. It is likely that the Cl^- ions will accelerate atmospheric corrosion in such cases.

It should also be expected that in corrosion processes influenced by gaseous chlorine, the hygroscopic nature of the chlorides produced in the reaction will lower the critical humidity of the system. This lengthens the periods of surface wetting, and so promotes corrosion. Again, the nature of the chloride is decisive here. Stable basic chlorides, such as those formed on zinc, copper and cadmium, reduce the Cl^- activity, so that the action is limited to the corrosion product-atmosphere interface. As already discussed, iron does not form sparingly soluble chlorides, and so the influence of corrosion products on iron does not reduce the activity of chloride ions acting at the iron–rust interface to promote the anodic part-process.

The stimulating effects of both gaseous hydrogen chloride and chlorine are stronger than that of Cl^- ions occurring as salts (e.g. NaCl) in the atmosphere.

The reason lies in the acid character of the two former species, which cannot be completely counteracted by the corrosion products and thus can accelerate the dissolution processes at the outer phase boundary, especially on non-ferrous metals such as zinc or copper. This effect should not be so pronounced on iron, since Heusler says that the rate of anodic dissolution of iron increases with increasing OH^- activity.

3.5.4   Ammonia

This is one of the more frequent atmospheric pollutants. Its role in atmospheric corrosion has been little investigated to date. It has a major effect on stress corrosion cracking of copper alloys, especially brass (season cracking). Ross and Callaghan have recently investigated the influence of ammonium ions on atmospheric rusting of steel. NH_4^+ ions e.g. as solutions of ammonium chloride or sulphate, liberate ammonium hydroxide solution from the outset of the corrosion [70]. This solution has a lower surface tension than water or salt solutions, and thus helps to wet the surface. This phenomenon is only significant at the initial stages of the rusting, however; its importance disappears in the presence of an already-coherent rust layer.

3.5.5   Solid matter

Examination of different investigations of the effects of dust on atmospheric corrosion [55, 56] leads to several basic conclusions:

Dust appears particularly influential only in the initial stages of atmospheric corrosion i.e. while there is still no perceptible formation of corrosion products. Thick corrosion product layers assume complete control of the corrosion process.

In the initial period of atmospheric corrosion in presence of solid material, the properties of the electrolyte formed (and hence the solubility of the material) are very important. Assuming that atmospheric impurities accelerate the corrosion mainly by promotion of the anodic reactions, solid matter whose solutions contain sulphate, chloride and similar anions must be seen as particularly active. It is well-known that the critical humidity is of extreme importance for atmospheric corrosion. It has already been discussed in Chapter 3.3 how the water vapour pressure over saturated salt solutions determines the value of this. The 'hygroscopicity' of the solid impurities deposited on the surface must therefore also be considered. Their soluble, electrolyte-forming component is thus of extreme importance.

The older viewpoint [57] that physically-produced water accumulation (by capillary condensation) may promote corrosion has not been confirmed. Solid materials with water uptake ability of this type, such as silica gel or active carbon (soot), do indeed take up considerable quantities of water, but the water is bound so strongly to the adsorbent that it is not available for electrolyte formation.

Similarly, the catalytic effect of soot on oxidation of SO_2 to sulphuric acid is always overemphasized. Though this action is observed in the initial stages of

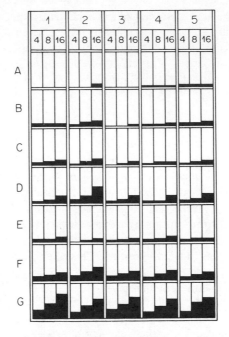

Figure 23. Influence of dust on the atmospheric corrosion of iron.
1. Blank test; 2. Silica dust; 3. Active carbon; 4. Glass dust; 5. Industrial dust. A. 30% relative humidity; B. 50% relative humidity; C. 70% relative humidity; D. 90% relative humidity; E. 30% relative humidity and 0.3 g SO_2 m^{-3}; F. 70% relative humidity and 0.3 g SO_2 m^{-3}; G. 90% relative humidity and 0.3 g SO_2 m^{-3}.
4, 8, and 16 are the lengths of the experiments in days.

corrosion at very high SO_2 concentrations, it can be shown that such small amounts of SO_2 as do reach the surface from the atmosphere are converted to SO_4^{2-} ions within a very short time even in the absence of soot. Already-existing corrosion products are sufficiently catalytic to convert all the SO_2 to sulphate (see Chapter 3.5.2).

Thus, solid matter (dust) deposited from the atmosphere onto metals may exert an accelerating influence on atmospheric corrosion. This is especially pronounced in the initial stages, because of the action of soluble electrolyte-forming components in reducing the critical humidity and providing activating anions which promote the anodic part-processes. If solid corrosion products are already present, the effect of them is lessened, possibly because of the arrival of fresh amounts of activating ions. The corrosion products provide the main controls on the process, however.

3.6 Properties of solid corrosion products and their influence on the course of atmospheric corrosion

In the discussions of the two part-processes of the electrochemical reaction which make up atmospheric corrosion, and especially in the section on the mechanisms of action of atmospheric impurities, attention was drawn to the mechanisms of formation of solid corrosion products, and to their properties. It seems appropriate to summarize data on solid corrosion products at this point.

The corrosion products formed as a new phase in the metal-atmosphere system undoubtedly influence the long-term course of atmospheric corrosion, and this is related to their properties and formation mechanisms. Examination of the anodic and cathodic reactions shows that the anodic path is responsible for their formation. Two mechanisms must be considered: in the first, soluble salts are formed as primary products, while in the second there is direct formation of sparingly soluble oxide or hydroxide compounds as the primary anodic reaction. An example of the first type of mechanism is the formation of ferrous sulphate during rusting of iron under atmospheric conditions [41].

$$Fe + H_2O \rightarrow Fe(OH^-)_{ads} + H^+ \qquad (3\text{-}37)$$

$$Fe(OH^-)_{ads} \rightarrow Fe(OH)_{ads} + e^- \qquad (3\text{-}38)$$

$$Fe(OH)_{ads} + SO_4^{2-} \rightarrow FeSO_4 + OH^- + e^- \qquad (3\text{-}39)$$

$$FeSO_4 \rightarrow Fe_2^+ + SO_4^{2-} \qquad (3\text{-}49)$$

An example of the second type is reactions (3-50) and (3-51):

$$Me + H_2O \rightarrow MeO + 2H^+ + 2e^- \qquad (3\text{-}50)$$

$$Me + 2H_2O \rightarrow Me(OH)_2 + 2H^+ + 2e^- \qquad (3\text{-}51)$$

Similar compounds are formed by reaction of anodically-produced metal ions and hydroxide ions (Equation (3-52)):

$$Me^{n+} + nOH^- \rightarrow Me(OH)_n \qquad (3\text{-}52)$$

There are many possibilities of further reversible and, particularly, irreversible conversions in the liquid water-oxide-hydroxide system.

When other reactive species, especially anions, are introduced to the basic metal-water-air system due to air impurity, the form of the system changes. These anions can, depending on their activity, participate directly in the anodic primary step or influence secondary conversion of the corrosion products. An example of the first case is the atmospheric corrosion of iron in a sulphate-containing electrolyte. Above a certain $SO_4^{2-}:OH^-$ activity ratio, direct formation of hydroxide rust components gives way to a faster reaction involving primary formation of sulphate (i.e. equations (3-37) to (3-39) and (3-49) replace equation (3-53)).

$$Fe + 2H_2O \rightarrow FeOOH + 3H^+ + 3e^- \tag{3-53}$$

(written as the overall reaction).

Oxidative hydrolysis can lead to the same end product:

$$Fe^{2+} + 2H_2O \rightarrow FeOOH + 3H^+ + e^- \tag{3-54}$$

It is typical of this type of reaction that the stimulating sulphate anions do not remain bound in the product as a sparingly soluble form, and so the activity of these ions falls off only slowly. This falling off is possibly caused by binding into the two most important rust components, lepidocrocite and goethite.

On metals which can form stable basic salts with sulphate and chloride, which are the two most important anions in promotion of atmospheric corrosion, these anions can participate directly in the primary anodic step. For example, Feitknecht found the following reactions to occur at suitable $Cl^- : OH^-$ ratios:

$$5Zn + 8OH^- + 2Cl^- \rightarrow ZnCl_2 . 4Zn(OH)_2 + 10e^- \tag{3-55}$$

$$7Zn + 12OH^- + 2Cl^- \rightarrow ZnCl_2 . 6Zn(OH)_2 + 14e^- \tag{3-56}$$

Both of these products are described as formed directly in the anodic reaction.

Similar compounds can also be formed in secondary reactions from primarily-formed oxides and hydroxides. Patina formation on copper may be described by the reaction:

$$4CuO + SO_2 + 3H_2O + \tfrac{1}{2}O_2 \rightarrow CuSO_4 . 3Cu(OH)_2 \tag{3-57}$$

The formation of basic salts on other metals can similarly be via primary or secondary reactions.

If the $SO_4^{2-} : OH^-$ or $Cl^- : OH^-$ ratio is especially high and the electrolyte is simultaneously temporarily acidic, soluble normal salts are formed as secondary products because the conditions for stability of the basic salts are absent. Depending on the partial pressure of HCl(g), the action of this gas on primarily-formed $Zn(OH)_2$ layers is by the reactions [53, 58]:

$$7Zn(OH)_2 + 2HCl \rightarrow ZnCl_2 . 6Zn(OH)_2 + 2H_2O \tag{3-58}$$

$$5Zn(OH)_2 + 2HCl \rightarrow ZnCl_2 . 4Zn(OH)_2 + 2H_2O \tag{3-59}$$

$$Zn(OH)_2 + 2HCl \rightarrow ZnCl_2 + 2H_2O \tag{3-60}$$

$Zn(OH)_2$ layers are also attacked by SO_2-containing atmospheres. In this case, and that of cadmium, crystallographically-defined basic sulphates can be identified, and so a good description of the reaction stoichiometry is possible. High sulphate contents (up to approximately 16%) are found in chemical analyses of zinc corrosion products from industrial regions [59], with only a

Table 7. Compounds found frequently in atmospheric corrosion products

Metal	Compounds	Notes
Iron	α-FeOOH, γ-FeOOH, β-FeOOH, Fe(OH)$_2$, Fe$_3$O$_4$, (Fe$_2$O$_3$?), aFeSO$_4$. bFe(OH)$_2$. cFe(OH)$_3$	β-FeOOH only in atmospheres containing chloride
	FeSO$_4$. 4H$_2$O, FeSO$_4$. 7H$_2$O	in industrial atmospheres
	FeCl$_2$	in marine atmospheres
Copper	CuO, Cu$_2$O, CuCO$_3$. 2Cu(OH)$_2$, 2CuCO$_3$. Cu(OH)$_2$	
	CuSO$_4$. 3Cu(OH)$_2$, CuSO$_4$. 2Cu(OH)$_2$, CuSO$_4$. 5H$_2$O	in industrial atmospheres
	Cu$_2$(OH)$_3$Cl	in marine atmospheres
	CuS	
Zinc	ZnO, ε-Zn(OH)$_2$, β-Zn(OH)$_2$, ZnCO$_3$, ZnCO$_3$. Zn(OH)$_2$	
	2ZnCO$_3$. 3Zn(OH)$_2$, ZnCO$_3$. 3Zn(OH)$_2$. H$_2$O	
	ZnCl$_2$. 4Zn(OH)$_2$, ZnCl$_2$. 6Zn(OH)$_2$	in marine atmospheres
	ZnSO$_4$. 4Zn(OH)$_2$, ZnSO$_4$. xH$_2$O	in industrial atmospheres
Cadmium	CdO, Cd(OH)$_2$, CdCO$_3$, 2CdCO$_3$. 3Cd(OH)$_2$	
	CdSO$_4$. H$_2$O, Basic sulphate (?)	in industrial atmospheres
	CdCl$_2$. H$_2$O, Cd(OH)Cl	in marine atmospheres
	CdS	
Aluminium	Al(OH)$_3$ gel, γ-Al$_2$O$_3$, γ-AlOOH, α-Al(OH)$_3$ amorphous basic sulphate	in extremely SO$_2$-polluted atmospheres
	amorphous basic chloride	in marine atmospheres
Lead	PbO, Pb(HCO$_3$)$_2$, PbSO$_4$, basic sulphate	
Magnesium	MgO, Mg(OH)$_2$, basic carbonate, MgSO$_4$, MgCl$_2$, basic chloride	

small part of the sulphate in soluble forms. There is relatively strong binding of the anions in the corrosion product in this case.

A review of the individual products of atmospheric corrosion is given in Table 7.

Both the chemical and physical properties of the corrosion products are important in determining the course of corrosion. It is obvious from Table 7 that almost all are crystallographically well-defined, though colloids are also possible. Especially in rust, diffraction methods show a significant amorphous component, whose composition may be described by the oxyhydroxide formula FeO$_n$(OH)$_{3-2n}$ [93]. This plays an important role in the formation of protective rust layers on low-alloy weathering steels. Its capacity to bind water is particularly important here, since it causes a slowing of the conversion to crystalline α-FeOOH, which does not protect the steel surface. Amorphous rust probably arises as a secondary precipitation product of primarily-formed γ-FeOOH which has dissolved in the surface electrolyte.

Alloying additions in weathering steels (Cu, Ni, P, Si, etc.) promote primary γ-FeOOH formation and 'stabilize' the amorphous rust against water loss and conversion to crystalline α-FeOOH. Local differential formation of corrosion products by direct anodic oxidation may sometimes lead to epitaxial phenomena; this has not yet been proven. The crystalline character of the corrosion products is closely related to their semi-conductor properties, the more so in that one expects a decrease in metal ion concentration in the corrosion products in moving towards the atmosphere-product interface, and a decreasing oxygen content in the opposite direction, in the metal-corrosion product-atmosphere redox system. Definite electron conductivity by rust has been demonstrated by e.g. Herzog by use of rust sheets as electrodes in electrochemical measurements [46]. Photo-effects at the zinc-corrosion product and copper-corrosion product interfaces, which are doubtless connected with the semi-conducting properties of the corrosion products, have also been reported [29, 60]. Particularly on zinc, it can be shown that the corrosion product conductivity is due to excess zinc ions in the lattice. The photo-effect can be enhanced by reduction and decreased by oxidative treatment [29].

The semi-conducting structure of the corrosion products can also be related to their catalytic action in oxidation and reduction processes. As already discussed (Chapters 3.5 and 3.5.2), it must be assumed that the following reactions, which are important in corrosion, are affected by semi-conductor properties of products:

$$O_2 \rightarrow 2O_{ads} \tag{3-17}$$

$$O_{ads} + 2e^- \rightarrow O^{2-} \tag{3-62}$$

$$SO_2 + O_2 + 2e^- \rightarrow SO_4^{2-} \tag{3-35}$$

$$SO_2 + 2O^{2-} \rightarrow SO_4^{2-} + 2e^- \tag{3-63}$$

A similar mechanism, aided by the presence of reaction-catalysing corrosion products, was described by Knotková [61] for the accelerating effect of formaldehyde on the atmospheric corrosion of zinc and other metals. Here, only formates were found in the corrosion products, and no formaldehyde.

The extent to which conversion (SO_2 to SO_4^{2-}, HCHO to HCOOH) is aided by homogeneous catalysis by dissolved metal ions in the surface electrolyte remains unclear. In each case, Knotková could show that the soluble component of the corrosion product (obtained by leaching out) accelerates the reaction $SO_2 \rightarrow SO_4^{2-}$ very strongly [44].

Other physical and chemical properties of the corrosion products and of the metal-corrosion product system should be examined more closely e.g. their adhesion, cohesion, ion exchange ability (especially of their colloidal component), grain size, etc. There is little data on these, though they are very useful properties for comprehensive discussions of atmospheric corrosion.

The investigation of corrosion-affecting properties of atmospheric rust is of particular technological significance. Rust is composed mainly of two chief components: lepidocrocite and goethite (α and γ-FeOOH). In addition to these, it invariably contains magnetite (Fe_3O_4), amorphous, usually hydroxide-based, components, and either ferrous sulphate ($FeSO_4 \cdot 4H_2O$) or ferrous chloride ($FeCl_2$), depending on the atmospheric impurities. The former arises by the conversion of atmospheric SO_2, the latter in chloride-containing atmospheres (in coastal areas) [37]. Analyses of rust indicate that in spite of the relatively high enrichment by SO_4^{2-} or Cl^- (2 to 3%), only a small fraction of the anion is water-soluble, though the apparent compound formulae are $FeSO_4$ and $FeCl_2$. Reviews of this have been given by e.g. Ross [62] and Honzák [63]. In the latter work, differential vertical rust compositions were sought. It is possible to discover three layers in rust. That on the surface is easily removable (e.g. by light scraping), the middle layer can be 'burst off' by bending the specimen, and that lying on the metal surface is very adherent and cannot be removed mechanically. Table 8 reviews this.

Table 8. Typical layering of rust deposits, and the compositions of the individual layers. (Rust grown for 34 months in an industrial atmosphere)

	Layer		
	I	II	III
Amount of rust (mg cm^{-2})	7	40	29
Iron content in rust (%)	60·5	63·9	66·9
Fe^{2+} (%)	0	8·66	60·5
SO$_4^{2-}$ (μg cm^{-2})	281	1128	240
Soluble SO$_4^{2-}$ (μg cm^{-2})	55	87	40
Cl$^-$ (μg cm^{-2})	32	186	130

Layer I: removed by brushing or scraping.
Layer II: 'sprung off' by bending the specimen.
Layer III: removed by chemical methods (e.g. pickling).

The highest anion content is found in the inner layers, which is in basic agreement with the x-ray results of Laengle [58] and the metallographic investigations of Ross, Callaghan, et al. [62]. Undoubtedly, these results are related to the earlier-discussed mechanism of $FeSO_4$ formation and the formation of 'sulphate nests'.

3.7 The kinetics of atmospheric corrosion

In this chapter, theories on the kinetics of the processes will be examined on the basis of the mechanisms of atmospheric corrosion discussed so far. Thus, only the narrow time-span in which the process is 'active' will be considered i.e. the period when sufficient electrolyte is present. (Chapter 3.8 deals with the long-term course of atmospheric corrosion, in which the electrolyte is alternatingly present/absent).

The rates of the different reaction mechanisms of atmospheric corrosion are influenced particularly by the following factors: the metal type; the properties of the atmosphere which affect corrosion; the corrosion products, which characterize the state of the metal–electrolyte–atmosphere system by their amount, chemistry, and phase composition, and exhibit more or less constant properties once the 'stationary' state has been reached. Since these properties often depend on the metal type and specific atmospheric properties, they should be considered alongside these two factors rather than in isolation.

3.7.1 Influence of metal type on atmospheric corrosion kinetics

The metal character must be examined with special respect to the chemical properties of the corrosion products formed and their formation mechanisms. For this reason, it is desirable to split the metals into four groups:

1. Rare metals, which remain immune in atmospheric surface electrolytes on thermodynamic grounds in the sense of the definition by Pourbaix [12]. In this group are the platinum metals, gold, and, under suitable conditions, those alloys with sufficient rare metal content.

2. Metals which achieve a passive state, which is not disturbed by the relatively low aggressiveness of atmospheric electrolytes so that they remain passive. Included in this group are chromium, titanium, zirconium, easily passivated steels with high chromium content, high chromium-content nickel alloys, and aluminium and some of its alloys. To a limited extent, antimony, molybdenum, tungsten, etc. could perhaps be included here, though they are considered more rarely from the viewpoint of their corrosion resistance. (Note: Silver perhaps belongs rather in the group with 'true' passivity in the atmosphere, since there is always a very thin layer of oxide on its surface, and it is easily attacked to form the sulphide. The corrosion resistance of silver is, however, obviously connected with the noble overpotential of its corrosion reactions, which permits only low reaction rates under atmospheric conditions).

General knowledge of the depassivation phenomena on these metals (e.g. [64]), especially on chromium-containing steels and aluminium, indicates that in particularly aggressive atmospheres with a high content of activating impurities (especially chloride, but also dust or sulphur compounds), the passivity may be disrupted locally, and e.g. pitting corrosion of varying intensities (though mostly superficial) may occur. Similarly, atmospheric electrolytes can promote a more or less strong grain boundary attack and stress corrosion cracking (both especially on aluminium alloys) by disrupting the inherent passivity.

Disruption of 'true passivity' in the atmosphere is thus caused by atmospheric aggressiveness, which is related in turn to its content of activating species.

3. In the third group are included those metals and alloys whose corrosion products are formed mainly by direct anodic hydroxide or oxide formation. This leads to formation of thicker layers than in 'true' passivity during the corrosion process. Their formation and disruption mechanisms are, however, related to those of Group 2. The oxide and hydroxide layers (and under certain conditions, also basic salts formed by direct anodic reaction), which are invariably only slightly water-soluble, are converted into soluble compounds at the surface by atmospheric impurities and their conversion products (see Chapter 3.4), and under suitable conditions will be washed away from the surface. This process is more intensive the greater the influx of aggressive ions, and active dissolution may set in if the process occurs more rapidly than does anodic

reformation of the hydroxide (oxide) layer. Such behaviour is found, for example, in railway tunnels [42], where the high SO_2 content may cause direct formation of $ZnSO_4 . 7H_2O$ as the predominant corrosion product. It can be shown similarly in the laboratory that at SO_2 concentrations of 0·5 to 1%, only crystalline copper sulphate, $CuSO_4 . 5H_2O$, is formed on copper.

The technically more important non-ferrous metals, particularly zinc, copper, cadmium, magnesium, nickel (which remains passive in little-polluted atmospheres), tin and lead, corrode in the atmosphere according to these mechanisms. Lead behaves differently from the others, however, in that its oxide corrosion products form slightly soluble normal salts in SO_2- (or SO_4^{2-}-) or Cl^--containing environments. ($PbSO_4$ is only very slightly soluble). This is the reason for the particularly high corrosion resistance of lead in industrial atmospheres with high sulphur dioxide content.

4. On its own in the fourth group is the metal used most in technology, iron (in the form of construction steel or cast iron). Following from the ideas of (3) above, the particularly large (approximately an order of magnitude greater) corrosion rate of this metal, particularly in aggressive industrial and marine atmospheres, can be described by invoking a partly-actively proceeding anodic dissolution. As was mentioned earlier in the discussion of the accelerating effects of atmospheric corrosion stimulators, their action on iron is unique in that they alter the reaction mechanism at the iron–rust boundary, and their accelerating conversion products (especially sulphate ions) are always re-liberated in consequent reactions. This makes the extremely poor corrosion resistance of iron easily understandable.

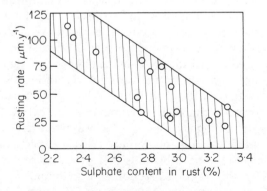

Figure 24. The relationship between rate of rusting and the sulphate content of the rust [67].

The so-called rust–bearing (weathering) steels contain alloying additions which chiefly enhance the stability of the amorphous rust [94]. However, the ability of the dissolution products of these additions to bind stimulator anions, especially sulphate, in insoluble forms also plays a definite role [65–67].

3.7.2 Influence of atmospheric properties on the corrosion kinetics

The properties of the atmosphere which are important for corrosion were discussed in Chapter 2. It is sought here to analyse their influence on the reaction kinetics, on the basis of the reaction mechanisms discussed earlier.

This can be based on a series of literature results (e.g. [68]) and on specialist publications on this topic. Continuing on from the above discussion of the effects of metal character, the reaction kinetics of the metals belonging to Groups 3 and 4 are specially considered.

The metals in the third group have corrosion mechanisms which may be represented schematically for formation of divalent compounds as:

$$Me + H_2O \rightarrow MeO + 2H^+ + 2e^- \tag{3-50}$$

$$Me + 2H_2O \rightarrow Me(OH)_2 + 2H^+ + 2e^- \tag{3-51}$$

$$MeO + 2H^+ \rightarrow Me^{2+} + H_2O \tag{3-64}$$

$$Me(OH)_2 + 2H^+ \rightarrow Me^{2+} + 2H_2O \tag{3-65}$$

$$MeX_2 + 2H_2O \rightarrow MeX_2 \text{ (solution)} \tag{3-66}$$

The first two of these equations describe the formation of corrosion products (mainly by direct oxidation at the anode), and the later ones their degradation by conversion into soluble salts ($X = Cl^-$ or SO_4^{2-}).

This simply-represented reaction mechanism leads to important conclusions regarding reaction kinetics: the formation of corrosion products should be accelerated by increasing water activity, and their breakdown retarded. Since in most cases the reactions which lead to disruption of the corrosion product layer are rate-determining and so proceed at the corrosion product-atmosphere interface, they should not be accelerated by increasing water activity if the concentration of atmospheric corrosion stimulator remains constant. This is demonstrated by investigations of the temperature dependence of atmospheric corrosion of zinc, copper and cadmium i.e. of metals which are typical representatives of this reaction mechanism. Experiments at constant relative humidity and constant atmospheric stimulator content (see Figure 22) show the undoubtedly inhibiting effect of p_{H_2O} (i.e. water activity) as it increases with temperature; this influence obviously outweighs the expected acceleration of the chemical reactions to form soluble salts [53]. This may be explained by promotion of hydrolysis in the opposite direction, or with the aid of primary steps which arise in concurrent adsorption of accelerating (Cl^- or SO_4^{2-}) and retarding (H_2O) species in these heterogeneous systems.

Using these theories, it seems clear why virtually no acceleration of non-ferrous metal corrosion (i.e. metals in Group 3) in damp, warm regions such as tropical rain forests is observed, and why relatively low corrosion rates of these metals are found in damp, warm industrial atmospheres.

Iron (the sole member of Group 4) behaves differently in this respect. It was suggested in the foregoing explanation of the differences between the mechanisms of atmospheric corrosion of iron and those metals in Group 3 that electrochemical reactions at the iron-rust interface determine the rusting, and that the kinetics of these reactions appear to be second order with respect to water activity and first order with respect to SO_4^{2-} activity (in SO_2-polluted atmospheres). For this reason, the two corrosion-influencing atmospheric factors (viz. water and stimulator content) do not act in opposite directions, as in the previous case, but both now accelerate rust formation. Similarly, temperature increase at constant relative humidity promotes rusting; this is closely related to the synergistic action of the two factors cited above and also to the specific reaction kinetics controlled by chemical processes. This is valid only to a limited extent: at very high temperatures (and hence high p_{H_2O}) e.g. at 40 °C, the inhibiting action of water vapour again begins to be significant [53] (Figure 25).

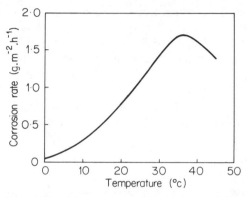

Figure 25. The temperature dependence of the
rate of atmospheric corrosion of iron.

Summarizing, then, three quantities of the atmosphere are found to have the most influence: water activity in the surface electrolyte, activities of accelerating anions (especially Cl^-, SO_4^{2-}) in the electrolyte, and temperature.

The activity of water in the surface electrolyte layer depends closely on its formation mechanism. As was examined more closely in Chapter 3.3.1, the transformation $H_2O_{(g)} \rightarrow H_2O_{(1)}$ occurs by sorption (or capillary condensation), particularly on colloidal components of corrosion products, and water uptake by water-soluble species. The conversion of water vapour into liquid water should therefore not be divorced from the concept of relative humidity. The results of Bartoň and Bartoňová [49] from measurements of humidity dependence of the rusting process (i.e. for setting up of 'stationary' rust layers) support this, at least for iron. Assuming that the reaction is second order in water activity, it can be shown that this activity can be expressed by a relative scale of humidities. Between the limits of the 'critical' relative humidity (approximately 80%) and the saturation humidity, the expression

$$v_{Kx} = v_{K96} \frac{(H_x - 80)^2}{(96 - 80)^2} \tag{3-67}$$

applies, where v_{K96} = corrosion rate at 96% relative humidity, which is used as the basis of the calculation, because of its high reliability (see Figure 19).

No similar algorithm exists for the Group 3 non-ferrous metals. It can be concluded from the mechanisms of formation and destruction of their corrosion products (Equations (3-50) to (3-66)) that, on the one hand, water activity promotes corrosion product formation, but on the other hand it washes away soluble salts and limits the action of corrosion-accelerating species. The humidity dependence of the corrosion rate is not quadratic in these cases, but is far flatter, and possibly linear.

The accelerating action of stimulating anions on the corrosion can be dealt with similarly. For both iron and the Group 3 non-ferrous metals, the reaction mechanisms for corrosion product formation and destruction suggest a linear dependence on ion activity. On non-ferrous metals, however, there is a definite competitive effect between activating ions and 'passivating' water, and so the ratio of activities of the reactants becomes critical. (As was discussed earlier, this ratio affects the corrosion course in that at higher temperatures, where the absolute humidity reaches significant values, the accelerating action of the stimulator is lessened. Consequently, the corrosion does not have a positive temperature dependence.)

An increase in temperature causes acceleration of all chemical reaction rates, according to the Arrhenius equation:

$$k_t = Z \exp\left(\frac{-E}{RT}\right) \tag{3-68}$$

where k_t = corrosion rate, E = activation energy, R = thermodynamic constant, T = absolute temperature, and Z = constant. This applies to atmospheric corrosion reactions, specifically for iron (see Figure 22), and sometimes for non-ferrous metals, though physical primary processes (e.g. competing sorption of stimulating and passivating reagents) may also be rate-determining here. In these cases, the water activity is related to absolute rather than relative humidities.

Equations (3-69) and (3-70) can be derived on kinetic principles for periods of active corrosion:

For Group 3 non-ferrous metals:

$$v_K = k_1 \cdot a_{H_2O} \cdot k_2 \cdot \frac{a_x}{a_{H_2O}} \cdot \exp\left(-\frac{k_3}{T}\right) \tag{3-69}$$

and for iron:

$$v_K = k_4 \cdot a_{H_2O}^2 \cdot a_x \cdot \exp\left(-\frac{k_5}{T}\right) \tag{3-70}$$

where v_K = corrosion rate, a_{H_2O} = water activity, a_x = activity of stimulating anion, which is related to the corresponding air impurity, and $k_{1 \text{ to } 5}$ are constants.

These relationships are valid after establishment of stationary corrosion rates, when the k values become constant; until then, the k values change with time.

3.8 The long-term course of corrosion

A knowledge of the relationships between the different data which allow characterization of atmospheric properties and long-term atmospheric corrosion courses is certainly useful. From this, it should be possible to derive rational protection methods and to derive a classification scheme of aggressiveness of typical atmospheres as a basis for choice of correct corrosion protection methods. This can then allow a system of reliable, authoritative and rapid test methods.

In this chapter, these relationships will be examined more closely. National data on meteorological measurements (which are published by the widespread and relatively dense network of stations, and have been complemented recently by air pollution measurements) and the ideas on kinetics and mechanisms of atmospheric corrosion discussed earlier may be combined to give a picture of the long-term course of corrosion as a function of meteorological factors and air purity.

Various data has been published in recent years and should be analysed here. Interesting results have come from Sweden [69]. Empirical equations, based on corrosion experiments on steel at different test stations, have been derived to describe the corrosion course in terms of atmospheric factors such as temperature, humidity, and air pollution. It is found that the temperature dependence of the process obeys Van't Hoff's rule. The temperature range of atmospheric corrosion is divided into 5° intervals, and the relative coefficients are calculated to be:

Temperature (°C):	0–5	5–10	10–15	15–20	20–25
Coefficient:	1	$2^{1/2}$	2	$2.2^{1/2}$	4

The influence of the relative humidity can be calculated similarly:

Rel. humidity (%):	100–90	90–80	80–70	70–60	60–50	50–40	<40
Coefficient:	1	0·93	0·78	0·53	0·17	0·03	0.00

The acceleration of corrosion by air pollutants is shown in Figure 26. The actual calculation is then done by multiplication of the hourly frequency of the particular meteorological measurement by the correct coefficient, addition of the results and correction for the air impurities. In this way, the 'meteorological' corrosion climate for Sweden was established and expressed as iso-pleths on the map.

This method i.e. the comparison of test results and meteorological or chemical data on the environment, is quite acceptable. It does not use the mechanism or kinetics of the corrosion process as the basis of its explanation of the long-term course of the process.

58

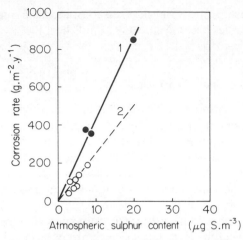

Figure 26. The relationship between the sulphur content of the atmosphere and the rate of steel corrosion.

Schwenk and Ternes [71] have published calculations based on rusting-time curves for unalloyed carbon steel and rust-bearing low-alloy steels. Four-year experiments in Duisberg and Gelsenkirchen gave corrosion-time relationships (Figure 27) which could be described by Equation (3-71):

$$i = k(t + t_0)^{-n} \tag{3-71}$$

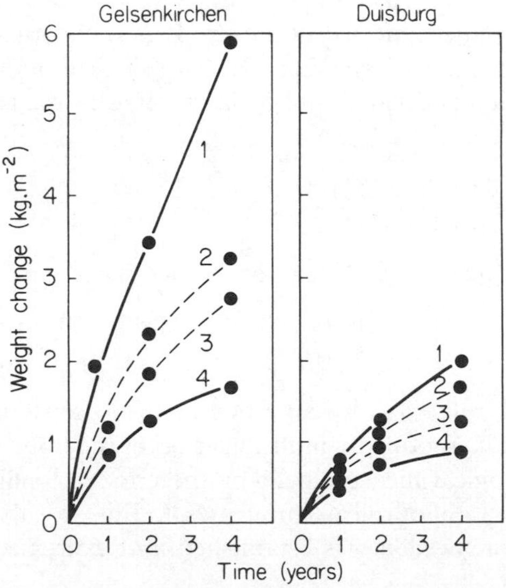

Figure 27. The dependence on copper content of rates of rusting of steels in different atmospheres [71].
1. 0·10% Cu; 2. 0·18% Cu; 3. 0·27% Cu; 4. 0·34% Cu.

where i = average corrosion rate in $kg\,m^{-2}\,yr^{-1}$, t = exposure time, t_0 = the period of around two weeks during initiation, when the relationship is invalid, and k and n are constants. This relationship holds for corrosion rates which decrease during the course of the reaction, because of protection of the surface by the rust which is forming. This leads eventually to a stationary corrosion rate [50], so that the corrosion-time curve becomes linear.

The most important idea of all, which runs through the work of Golubjev [72], Guttman, Sereda [73], Berukschtis, Klark [74] and Bartoň and Bartoňová [41], is the undoubtedly correct assumption that atmospheric corrosion is a discontinuous process which proceeds only when the surface electrolyte exists. These periods are related to the changing humidity and temperature conditions in the atmosphere. The total corrosion over a long period of time is thus the sum of corrosion effects during shorter periods.

$$K = \sum_{i=1}^{n} \tau_i \cdot v_{Ki} \tag{3-72}$$

where K = total corrosion loss, τ_i = duration of ith period of electrolyte presence, and v_{Ki} = average corrosion rate during the period τ_i.

The chief problem in setting up laws for the long-term course of atmospheric corrosion is thus determination of the relationships which define the values of v_K and τ as explicit functions of the properties of the metal and measurable meteorological variables.

The duration of electrolyte presence can be measured relatively accurately. One method involves use of a copper-iron or platinum-zinc macrocorrosion cell as a current generator [75, 76]. These cells give a sufficiently high current in the presence of surface electrolyte layers for measurement using a recorder. Using this method, Sereda [77] has measured durations of electrolyte presence in different typical atmospheres, and combined the results with measured values of air pollution by sulphur dioxide (as cumulative values of SO_2 coming into contact with the surface) into an empirical expression which describes the rate of corrosion of steel as a function of temperature and SO_2 concentration per day of electrolyte presence:

$$\log y = 0.130x + 0.0180z + 0.787 \tag{3-73}$$

(where x = temperature in °F, z = SO_2 adsorption in $mg\,dm^{-2}\,day^{-1}$, and y = corrosion rate in $mg\,dm^{-2}\,day^{-1}$). This equation is valid only for the initial stages of rusting, because of, on the one hand, the experiment conditions (in which the samples are corroded only for a short period e.g. a month) and, on the other hand, the frequently-confirmed nonlinearity of the corrosion-time relationship (see e.g. [50]). Another investigation [73] confirmed that, if the iron is covered with rust, the atmospheric conditions are no longer the only significant influence; the properties of the corrosion products now assume the dominant role. This is true for other metals besides iron e.g. zinc, copper, etc.

Berukschtis and Klark [74] similarly use the overall expression mentioned above for atmospheric corrosion, with closer definitions of v_K and τ to allow a more explicit expression. Besides measuring times of wetness with the macro-corrosion cells described above, they also used calculations of likely periods from meteorological data based on duration of precipitation (from pluviographic measurements). To these times were added the times needed for drying out of the electrolyte layers formed by precipitation, which were calculated from equation (3-74):

$$\tau' = \frac{\delta}{v} = \left(\frac{\delta}{v_0 - (75 - r)\, dv/dr + (20 - T)\, dv/dT} \right) n \qquad (3\text{-}74)$$

where δ = electrolyte layer thickness, v = vaporization rate in $\mu m\, h^{-1}$, v_0 = vaporization rate at $20\,^{\circ}C$ and 75% relative humidity = $35.5\,\mu m\, h^{-1}$, $dv/dr = -0.38$ per cent relative humidity, $dv/dT = +0.85$ per $^{\circ}C$, and n = number of periods. A constant thickness of the electrolyte layer ($100\,\mu m$) is obviously assumed here. The differential coefficients are determined empirically.

Because of the difficulty of access to pluviographic data, this equation is based on the assumption that electrolyte layers are formed exclusively by precipitation (rain, fog, melting snow), which contradicts the results of many authors (e.g. [41, 73]). Consequently, the equation can hardly be applied generally.

It is certainly true that a surface electrolyte is formed not only by vertical precipitation, but also by the direct conversion $H_2O_{(g)} \rightarrow H_2O_{(l)}$ on the corroding surface. This conversion occurs significantly below the saturation vapour pressure of water at the particular temperature. The critical humidity for

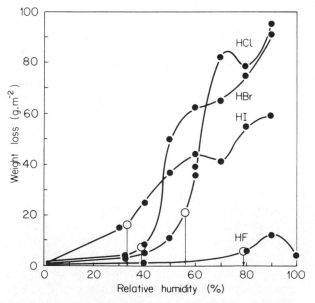

Figure 28. Critical humidities in the presence of different stimulators of corrosion [58]. (Critical humidity is indicated by open circle.)

atmospheric corrosion was defined forty years ago by Vernon [57]. Patterson and Hebbs [78] attempted to explain this in terms of capillary condensation onto the colloidal component of the corrosion products. More recent work confirms clearly that the electrolyte layers needed for an electrochemical reaction arise at relatively low humidities. Different investigations have shown that the value of 'critical humidity' is determined by the water vapour pressure over a solution of the soluble materials present on the corroding surface [58]. These materials occur as the products of reactions of the metal with chemically-active atmospheric impurities, or directly as soluble solid salts precipitated from the environment. These concepts have been basically confirmed by e.g. Buckowiecki [17], Preston [19], and particularly Laengle [58]. Kaesche [15] also concluded that the hygroscopicity of the soluble species on the surface is decisive in the formation of electrolyte layers, and that capillary condensation may be neglected as less important.

Atmospheric rusting of steel is undoubtedly the most important problem in atmospheric corrosion. Chapter 3.5 has already closely considered the mechanism of acceleration by atmospheric impurities, especially sulphur dioxide. From the 'critical humidity' point of view, rusting in industrial and urban atmospheres may be defined as an $Fe—FeSO_4—H_2O_{(g)}—O_2$ system. Determination of the kinetics of water uptake and corrosion in this system indicates that the critical humidity for it always lies below the relative humidity over a saturated $FeSO_4$ solution. This may be due to oxidative hydrolysis of $FeSO_4$, in which sulphuric acid, which can take up more water, is formed, or to the capillary condensation of water vapour onto colloidal hydroxide corrosion products, as suggested by Patterson and Hebbs [78]. An analogous experiment examining the humidity dependence of rusting in absence of hygroscopic salts

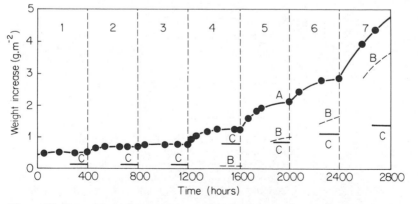

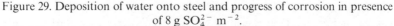

Figure 29. Deposition of water onto steel and progress of corrosion in presence of $8 \text{ g SO}_4^{2-} \text{ m}^{-2}$.
1. 21·5% relative humidity; 2. 35% relative humidity; 3. 60% relative humidity; 4. 72·5% relative humidity; 5. 87% relative humidity; 6. 96% relative humidity; 7. 100% relative humidity.
A. Total weight increase; B. Weight increase due to formation of corrosion products; C. Possible weight increase due to uptake of water of crystallization.

but presence of stimulating anions (Figures 29 and 30) suggests the latter of these two alternatives [80]. It is also found in these cases that corrosion begins only at an 'above-critical' relative humidity, and the corrosion rate is very low. Both plots show the critical relative humidity for atmospheric rusting to be 80 %. The higher water uptake in the presence of sulphate ions does not correlate with the expected values from hydration of ferrous sulphate. At lower humidities, uptake is higher than expected, and when the corrosion is proceeding actively,

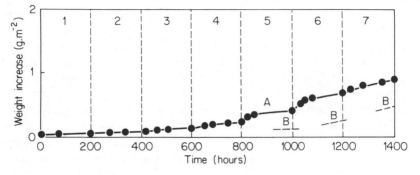

Figure 30. As Figure 29, in absence of stimulator.

lower (Figure 31). These phenomena can be explained by inhibition of the conversion of primary iron oxide gel into crystalline compounds (goethite, lepidocrocite) by the sulphate anions. The higher fraction of colloidal rust components favours physical uptake of water.

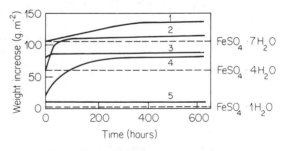

Figure 31. Uptake of water by $FeSO_4 \cdot H_2O$ at different atmospheric humidities.
1. 100 % relative humidity; 2. 96 % relative humidity; 3. 87 % relative humidity; 4. 72·5 % relative humidity; 5. 25 and 35 % relative humidity.

If it is further taken into consideration that at times when precipitation wets the surface, the humidity is typically higher than the critical value, it is possible to use the periods with 'super-critical' humidity to determine the time of wetness. This interval can be calculated with sufficient accuracy using meteorological data. Following a suggestion by Hudson [81], Bartoň and

Bartoňová have attempted to assess the time of wetness from meteorological measurements at three Czechoslovakian atmospheric corrosion testing stations [41]. The graphical evaluation of humidity and temperature data gave results like Figure 6. It is clear from the choice of coordinates (months versus hours) that sufficient measurements at periodic short intervals must be available. There is optimum reliability when hourly values are available. However, the normally-measured values will suffice; most meteorological stations around the world use 6-hour periods.

Some attention to the temperature data is required in interpreting the humidity diagram. At temperatures below the freezing point (the more concentrated the electrolyte solution, the lower its freezing point), the electrolyte freezes, and the corrosion becomes dormant. The influence of the temperature on 'super-critical' humidities must also be taken into account in calculating reaction kinetics of atmospheric corrosion.

Other methods for the calculation of the duration of electrolyte presence from meteorological data could obviously be found. The simplicity of the described graphical method is attractive, however. The greatest difficulty in this method is the absence of definite critical humidity values for the different metals. A practically useful limit value of critical humidity on iron may be assumed as $>80\%$. It is highly likely that this value is also valid for most non-ferrous metals. Meteorological data can therefore provide the means of determining individual and cumulative wetting times (τ in Equation (3-72)).

It is far more difficult to determine relationships between atmospheric data affecting corrosion and the corrosion rates of the different metals during the periods of wetting. Berukschtis and Klark [74] report an investigation which shows a functional connection between atmospheric data and corrosion rates during the period of wetting. These authors use the Taylor series to develop the implicit function

$$v_K = f(t, c) \tag{3-75}$$

into

$$v_K = K_0 + \frac{dv_K}{dt}\Delta t + \frac{dv_K}{dc}\Delta c + \frac{d^2 v_K}{dt \cdot dc}\Delta tc \tag{3-76}$$

where t = temperature in °C, c = SO_2 concentration in the surface electrolyte in mg l^{-1}, K_0 = corrosion rate at 20 °C and $c = 0$, and the differential coefficients are determined empirically.

There is no agreement between this relationship and the hypotheses on the mechanisms and kinetic effects of the two factors considered here (temperature and atmospheric impurity) which were presented in Chapters 3.5 to 3.7. These authors assume that sulphur dioxide dissolves in the electrolyte layer according to Henry's law. It is certain, however, that this gas is converted to sulphate extremely rapidly after its adsorption, and that, especially on iron, the whole of the amount of sulphur dioxide reaching the surface occurs as sulphate

[44]. Henry's law can therefore hardly be used to interpret the process. The equation neglects the humidity dependence of the process which was discussed earlier in this chapter. This is obviously related to the assumption of these authors, quoted earlier, that the corrosion proceeds actively only during and after precipitation periods when the excess of water is very great. It should be noted here that this equation was derived from short-term experiments (lasting about two months), and thus describes only the initial state of the corrosion process. Involvement of protective action by the corrosion products as they form was allowed for by the introduction of empirical, time-dependent coefficients.

It is obvious that the problem is chiefly the changing of the corrosion rate with time during the corrosion process. The more or less solidly adherent corrosion products which are formed retard the process, until a stationary corrosion rate is finally observed. An analysis of the generalized relationships between atmospheric corrosion and time enables the division shown in Figure 32 [82].

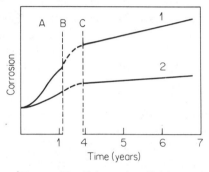

Figure 32. Schematic division of atmospheric corrosion.
A to B. Induction period; B to C. Establishment of stationary conditions; C. Stationary period.

1. Iron; 2. Non-ferrous metals.

Induction period: When technically pure metals are exposed to atmospheric influences, they always become covered with an 'inherent' oxide layer. Depending on the aggressiveness of the atmosphere, this layer affords a shorter or longer protection, until it is disrupted locally (as on iron) or over the surface (as on zinc or cadmium) and forms new corrosion products. The corrosion-time curves show deceleration during this period.

Establishment of the 'stationary' corrosion rate: after coverage of the whole metal surface with corrosion products, stationary conditions in the metal-corrosion product-atmosphere system are gradually established. Eventually, the corrosion products exhibit constant properties, such as thickness, chemical composition, and physical properties. After this state is reached, the process continues with a 'stationary' rate, which is characterized by a linear corrosion-time dependence. The assumption of a 'total' corrosion occurring over the individual corrosion-favoring periods is also valid here. The regulatory ability of the corrosion products ensures that the total corrosion is linear.

Basically, the two initial periods of the long-term atmospheric corrosion are shorter, the more aggressive the corrosion conditions. For example, in atmospheric rusting of steel the induction period lasts for years in very mild atmospheres (air-conditioned or heated rooms); in pure, humid outdoor atmospheres the stationary state is only achieved after several years, but is already attained in strongly polluted industrial regions after a few months (Figure 33).

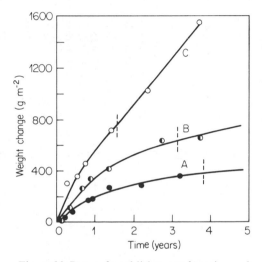

Figure 33. Rates of establishment of 'stationary' corrosion rates on steel during atmospheric corrosion in different atmospheres.

C. Aggressive industrial atmosphere.

A deeper insight into the processes which produce stationary corrosion rates on steels has been made possible by corrosion experiments in different types of aggressive atmosphere in Czechoslovakia. The meteorological characteristics (number of hours with relative humidity $> 80\%$ and temperature $> 0\,^\circ$C) of three test stations for which values of atmospheric pollution (in g SO_2 m^{-2} time^{-1}) are also available are given in Table 9. Progressive rusting of carbon steel was investigated during one of the years (1966) for which data is given in Table 9, at each of the three sites. The whole year was divided into 6-week periods during which separate groups of specimens were investigated for corrosion loss. Comparison of the corrosion-time curves for undisturbed growing layers and the short-term values added together as they are collected allows an assessment of the influence of the rust layer formed (Figures 34 to 36).

Table 9. Meteorological data for three test stations*

Year	Month	Hours with >80% relative humidity			T, °C			SO$_2$ g m^{-2} d^{-1}			SO$_2$ g m^{-2}		
		H	L	U	H	L	U	H	L	U	H	L	U
1965	1	435	0	744	0–1	–2	2	0.007	0.104	0.157	0.22	3.22	4.87
	2	54	432	0	1	0–1	–3	0.007	0.104	0.157	0.20	2.91	4.40
	3	352	300	354	3	0–1	2	0.004	0.070	0.153	0.12	2.17	4.74
	4	282	502	228	7.5	7	6	0.004	0.040	0.076	0.12	1.20	2.28
	5	303	430	253	10	10	14	0.004	0.040	0.076	0.12	1.24	2.36
	6	323	428	254	17	13	14	0.004	0.030	0.054	0.12	1.08	1.68
	7	340	456	266	17.5	14	17.5	0.004	0.020	0.096	0.12	0.62	2.97
	8	352	442	278	14	14	13.5	0.003	0.026	0.074	0.09	0.81	2.29
	9	378	533	288	12	10	10	0.003	0.037	0.074	0.09	1.11	2.22
	10	445	531	420	8	5	6.5	0.005	0.062	0.083	0.15	1.93	2.58
	11	336	720	509	4	3	3	0.007	0.080	0.083	0.21	2.40	2.49
	12	668	742	744	2	5	1	0.007	0.094	0.138	0.22	2.92	4.28
1966	1	0	0	0	0	–3	–5	0.029	0.107	0.133	0.86	3.31	4.12
	2	174	283	54	0–1	0	0	0.019	0.088	0.133	0.53	2.46	3.72
	3	437	692	312	3	4	3	0.009	0.071	0.131	0.28	2.20	4.05
	4	348	418	266	5	5	5	0.008	0.053	0.108	0.24	1.59	3.24
	5	319	335	156	11	10	10	0.006	0.038	0.071	0.19	1.21	2.20
	6	324	304	7	13	12	15	0.006	0.027	0.056	0.18	0.71	1.69
	7	334	314	43	17	15	17	0.006	0.019	0.051	0.19	0.57	1.58
	8	348	350	289	17	15	15	0.006	0.017	0.038	0.19	0.51	1.21
	9	389	389	360	14	13	14	0.006	0.021	0.052	0.18	0.63	1.56
	10	430	492	392	12	12	10	0.010	0.028	0.064	0.31	0.85	1.98
	11	477	705	417	7	5	5	0.020	0.037	0.059	0.60	1.11	1.77
	12	715	744	475	3	3	2	0.028	0.047	0.076	0.87	1.46	2.36
Σ 1965–1966		8553	10502	7083	Tτ† 8.1	7.5	7.5 °C	Σ 1965–1966			6.40	38.11	68.66

* H = Hurbanovo (pure rural atmosphere).   U = Ústí nach Labem (polluted chemical industrial atmosphere).
L = Praha Letňany (urban atmosphere).   † Tτ = average annual temperature for periods with relative humidity >80%.

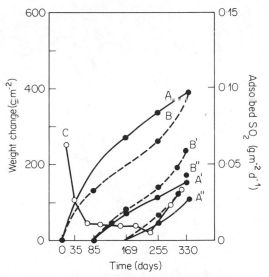

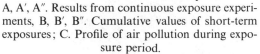

Figure 34. Influence of rust layers on progress of corrosion of steel in rural atmospheres.
A, A′, A″. Results from continuous exposure experiments, B, B′, B″. Cumulative values of short-term exposures; C. Profile of air pollution during exposure period.

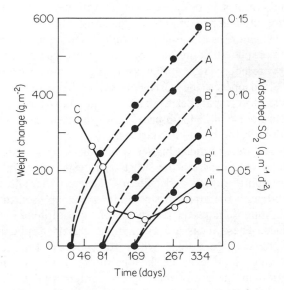

Figure 35. As Figure 34, for an urban atmosphere.

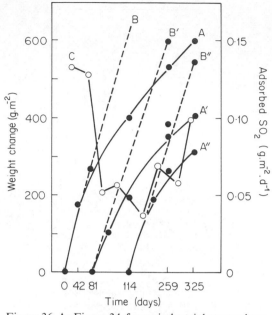

Figure 36. As Figure 34, for an industrial atmosphere.

Pronounced differences are found between the courses of corrosion on samples in very pure atmospheres in winter and summer. Experiments begun in winter, with high atmospheric sulphur dioxide content, have a gradual slowing-down of their corrosion rates during summer, and only reach the stationary value found on samples initially exposed in summer, which exists from the start of exposure, after approximately 12 months. The amount of rust formed during the summer months on samples initially exposed in summer is, however, not capable of completely counteracting the stimulating action of the greater amounts of SO_2 present in late summer, and the corrosion rate always increases. In these pure atmospheres, the corrosion values from uninterrupted growth of the rust layer closely parallel the sum of the short-term exposure values. This is thus a case of rusting at sub-critical SO_2 concentrations (in the sense of the ideas proposed in the previous chapter).

A different picture is found at higher atmospheric SO_2 contents (Figure 35). Here, the cumulative value from the short-term tests is invariably higher than the value from uninterrupted experiments. The rust here thus protects the underlying metal, with the attained stationary state representing the degree of protection. This state is characterized by a constant sulphate ion activity at the reaction interface. These ideas are even more clearly applicable to strongly polluted atmospheres (Figure 36), in which the summed short-term values give a straight line, and where, even at the beginning of the rusting process, the annual fluctuation in the atmospheric SO_2 content makes virtually no difference.

The phase of decreasing corrosion rate before the onset of the final 'stationary' period is thus connected with the formation of a corrosion product layer which, from a long-term point of view, exhibits constant properties. It seems important to emphasize that the concept of 'constant properties' must be so interpreted that the corrosion products are capable of counteracting seasonal changes in the corrosion conditions over long periods. (This applies especially to the effects of occasionally higher amounts of atmospheric impurities.) This can be seen in the experimental results discussed above, for observations after formation of a more or less 'protective' rust layer.

Similar behaviour should clearly be expected from non-ferrous metals, especially those which were included in Group 3 in Chapter 3.7.1 (i.e. zinc, copper, cadmium, etc.). The times required for establishment of 'stationary' corrosion rates in these cases are mostly shorter; this is related to the fundamentally different mechanisms of the rate-determining reactions. These periods are very short for zinc and cadmium (Figures 37 and 38).

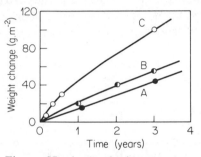

Figure 37. Atmospheric corrosion rates for copper in different atmospheres.
A. Rural atmosphere; B. Urban atmosphere; C. Aggressive industrial atmosphere.

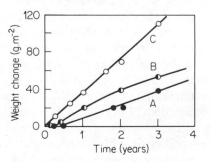

Figure 38. Atmospheric corrosion rates for zinc in different atmospheres.
A. Rural atmosphere; B. Urban atmosphere; C. Aggressive industrial atmosphere.

Stationary atmospheric corrosion: As was mentioned earlier, establishment of a long-term 'stationary' corrosion rate is the result of the corrosion product layer assuming regular properties. The corrosion-time curve is linear once this state has been reached, so that an analysis of the relationship between the curve gradient in this region and the meteorological and chemical data on the atmosphere is more easily performed than for the case of the changing corrosion rate during the initiation period.

Bartoň and Bartoňová have performed such an analysis [59]. Since atmospheric corrosion may be regarded as a discontinuous process (e.g. see Equation (3-72)), these authors have attempted to calculate the quasi-stationary corrosion rate i.e. to find an expression which allows determination of the corrosion rates in the short periods of active corrosion (at above-critical humidities), and to bring the relevant atmospheric variables into the relationship if possible.

As was described above, the times of active corrosion of steel, copper, and zinc in already-achieved stationary states were ascertained from the meteorological data summarized in Table 9 and the corrosion paths were determined from Figures 33, 37 and 38, for a two-year period. The corrosion rates calculated from these measurements are summarized in Table 10.

Table 10. Average stationary corrosion rates for iron, zinc and copper during periods of electrolyte presence (1965 to 1966)

Station	Fe $g\,m^{-2}\,h^{-1}$	Zn $g\,m^{-2}\,h^{-1}$	Cu $g\,m^{-2}\,h^{-1}$
Hurbanovo	0·0257	0·0033	0·0037
Letňany	0·0332	0·0032	0·0037
Ústí n. L.	0·1110	0·0108	0·0105

Since the calculated corrosion rates relate to uniformly-defined conditions (hours with relative humidity $>80\%$ and temperature $>0\,°C$), and the average temperatures at the three test stations are comparable, the different 'stationary' corrosion rates are due mainly to the effect of atmospheric sulphur dioxide content. A graphical treatment of this data for iron shows that the corrosion rate-sulphur dioxide (as the total amount of the gas reaching the surface) dependence is not linear. The test station with the purest atmosphere (Hurbanovo) had an atmospheric SO_2 content which appears less than the limit value required for acceleration of corrosion processes on iron. In Letňany, this value was narrowly exceeded. The critical value of SO_2 pollution of the atmosphere lies even higher for the non-ferrous metals, zinc and copper.

These limit values represent an annual average of 40 to 50 mg SO_2 m^{-2} day^{-1}, or calculated in volume units (according to Guttman [83] and Marek [84]; see Figure 39), a yearly mean value of 0·015 ppm.

Linear dependence of the corrosion rate on stimulator content of the atmosphere should be expected if this limit value is exceeded. The validity of

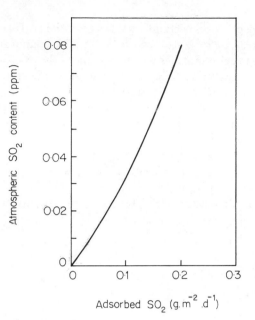

Figure 39. Relationship between the SO_2 content of the atmosphere and the amount of SO_2 deposited on a surface [7]. (An atmospheric SO_2 concentration of 1 ppm is equivalent to 2·86 mg SO_2 m^{-3} air.)

this idea is shown not only by theoretical studies (Chapters 3.5 to 3.7), but also by results of investigations under atmospheric conditions. Examples of these are the studies by Hudson and Stanners [85], Odén [69], Sereda [13], and Gutmann and Sereda [73] (Figure 40).

The goal of expressing the long-term course of atmospheric corrosion as a function of the metal type and the meteorological and chemical properties of

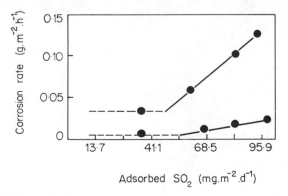

Figure 40. Dependence of rate of atmospheric corrosion on rate of uptake of SO_2 uptake onto the surface. Upper; iron; lower, zinc.

the different atmospheres is certainly attractive. If the corrosion aggressiveness of a particular area could be deduced from suitable manipulation of climatographic data, and other quantities which influenced the microclimate where the product was to be used were taken into account, the corrosion protection, in the true sense of the word, could be optimized.

This idea has been constantly re-examined in recent years, despite the continually pessimistic view of Stanners [94], who says there are two courses which can be followed:

1. An empirical course (e.g. [95, 96]), which will produce a functional relationship from computer-performed comparisons of corrosion values and climatic (or air chemistry) data.

2. A theoretically based course, based on the discussion given earlier on the discontinuity of the corrosion processes [63, 97, 98].

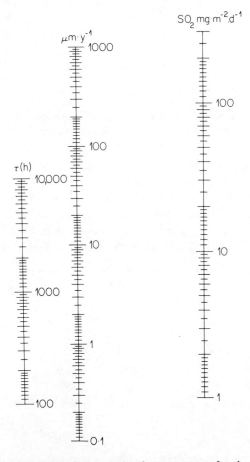

Figure 41. An example of a nomogram for the assessment of the 'stationary' corrosion rate of carbon steel in the atmosphere.

Barton and his coworkers have described a method which combines both of these courses [99]. They showed that long-term humidity and temperature data collection at a specific site allowed the estimation of a typical corrosion parameter τ (where τ expressed the long-term average value of the daily wetting time e.g. on the criteria discussed earlier, hours per day with relative humidity $\geq 80\%$ and temperature $\geq 0\,°C$). This value was obtained either graphically (from Figures 5 and 6 in Chapter 2.3) or by calculation from prolonged sets of daily data on humidity and temperature, which are published by the dense network of meteorological stations usually for 0700, 1400, 2100, and 0000 hrs. A third possibility is assessment of monthly τ values (i.e. hours per month) from an empirically derived stochastic relationship between the rainfall and moisture data for the month and the month's average temperature and humidity values, taken from meteorological measurements [100].

This defines one of the important parameters. The chemical-microclimatic influence of air pollutants can also be taken into account in the corrosion process; since SO_2 is the most widespread pollutant, this can be expressed in terms of $mg\ SO_2\ m^{-2}\ day^{-1}$. There is again a stochastic relationship between these variables and the SO_2 concentration measured in the environment.

The corrosion rate rate v_K (in $\mu m\ year^{-1}$) may be expressed as follows, in terms of these variables:

$$v_K = M \cdot \tau^n \cdot S^m \tag{3-77}$$

where M, n and m are constants which are determined by computer analysis of large sets of data on long-term measurements and results of atmospheric corrosion tests. These constants may be expressed in a nomograph (e.g. Figure 41). The constant M involves the specific corrosion kinetics of the metal concerned.

4
Non-uniform and Structural
Corrosion Phenomena

It was tacitly assumed in Chapter 3, on the chemical basis of atmospheric corrosion, that the only corrosion phenomena caused by the atmosphere as a corrosive environment were uniform across the surface. This is not true. There are important alloys on which the attack of atmospheric surface electrolytes is non-uniform, and produces e.g. pitting, intergranular corrosion or stress corrosion cracking.

4.1 Pitting corrosion

Even rusting, which is the best-known atmospheric corrosion manifestation, does not occur completely uniformly. If salt nests are formed during the rusting process, as is the rule in polluted industrial atmospheres or salt-laden coastal atmospheres, the salt deposits on the surface always leave behind individual pits of greater or lesser depth.

Even on easily-passivated alloys, typical pitting phenomena are not uncommon under atmospheric conditions, though the presence of these pits are not as serious a danger as is the presence of an aggressive environment. Though rust spots form over quite shallow pits on some 'stainless' steels, especially on martensitic or ferritic chromium-containing types, in strongly attacking environments these are mostly of consequence only from the point of view of the decorative aspect of the surface. The pit depth in such cases rarely exceeds 0·1 mm, and thus poses no threat to the strength of the material, the more so since the pits are uniformly distributed over the surface and do not pose problems from notch formation. The same applies to atmospheric corrosion of aluminium and its alloys.

The many theories of pitting corrosion will not be examined further here; they are described exhaustively in the literature [1, 2]. Activating anions which have been deposited on the surface are marked promoters of pitting corrosion initiation in the atmosphere; chloride and sulphate are particularly active. Carbon-containing dust particles also promote formation of corrosion pits.

It is typical in pitting corrosion of passive metals to find repassivation of part of the activated area and simultaneous formation of new active corrosion sites [2]. The temporary nature of the surface electrolyte and the invariably-present excess of atmospheric oxygen obviously aid the repassivation, so that it is easily understood why, even after many years, no deep corrosion pits are found.

The mechanism of pitting corrosion of decorative Cu–Ni–Cr electroplated layers on iron, zinc and aluminium alloys, which is a special case because of its impairment of the function of the layer, will be dealt with in detail in Chapter 5.5.6.

Washing of the surface by rain or by an artificial process reduces pitting corrosion, since it removes the stimulators/initiators. Pits are therefore always found to be deeper and more densely distributed on samples exposed to free air but protected from rain than on those exposed to the open air.

4.2 Grain boundary corrosion

'Pure' grain boundary corrosion in the atmosphere over extensive surface areas, independent of tensile stress, must be differentiated from the strongly localized stress corrosion cracking which often also proceeds intergranularly.

Technical alloys of base metals such as aluminium, zinc and magnesium are particularly prone to types of intergranular attack under certain conditions in aggressive atmospheres. Impurities and deliberately-added alloying additions of noble metals, which are known to allow formation of intermetallic phases during heat treatment which improve the mechanical properties of the alloy, are very important in these corrosion phenomena. Intermetallic compounds segregate particularly at grain boundaries, which leads to concentration gradients in the alloy composition. These gradients of the noble addition in the less noble bulk metal lead to local cell formation in the presence of a sufficiently conducting surface electrolyte. In this cell, the lowest alloying addition concentration is found in the bulk metal lying adjacent to the grain boundary and to the noble (i.e. cathodic) intermetallic phase [3]. In the simple case of a two-component alloy (e.g. Al–Cu), three regions with different potentials occur in the area around the grain. The most noble is the intermetallic phase (e.g. $CuAl_2$); the adjacent solid solution of copper in aluminium has the lowest copper content, so that both the intermetallic phase and the actual grain material act as cathodes. These theories of Akimov and Tomashov [4] can also be used for other two- and multi-component systems, provided metallographic analyses are available. Modern analytical methods, such as the microprobe, allow quantitative determination of concentration gradients even within grains, so that the likely effects of different treatment methods (e.g. heat treatment, etc.) to produce desired physical/mechanical properties on the likelihood of grain boundary corrosion susceptibility may be assessed.

It may be concluded from the theory of Akimov that there will be particularly injurious effects from those electrochemically noble metals which have only a very limited solubility in the less noble bulk metal. Thus, iron is the most dangerous impurity in alloys of aluminium, zinc and magnesium. Copper is less damaging in this respect, since it can dissolve up to relatively high contents, especially in aluminium and zinc.

Addition of base (i.e. non-noble) metals to aluminium as alloying agents poses little danger from the point of view of 'pure' intergranular corrosion, but is not totally innocuous. Intermetallic phases produced by different heat treatment methods or by aging can exist as either anodic or cathodic inclusions with respect to the bulk metal. The deficient zones, which adjoin the inclusions, have the greatest difference in potential from them. Grain boundary corrosion occurs on such multi-phase alloys according to the potentials established on the individual phases, and their distribution in the grains and at the grain boundaries. When, for example, the intermetallic phase Mg_2Al_3 acts as anode against the bulk metal (a solid solution of magnesium in aluminium) in an Al–Mg alloy and is segregated predominantly at the grain boundaries, intergranular corrosion proceeds by anodic dissolution of this intermetallic compound. If this phase is uniformly distributed, however (and this can be ensured by heat treatment), the danger of grain boundary corrosion is much less. Similarly, anodic phases such as $MgZn_2$ or $Mg_3Zn_2Al_2$, which form during

Table 11. Stationary electrode potentials of aluminium alloy phases, measured in a solution containing $53 \, g \, l^{-1}$ NaCl and $3 \, g \, l^{-1}$ H_2O at 25 °C [5]

Phase	Potential (V) versus 0·1 N calomel electrode
Mg	−1·73
Zn	−1·10
Mg_2Al_3	−1·07
Al + 4$MgZn_2$	−1·07
Al + 4Zn	−1·05
$MgZn_2$	−1·05
$CuMgAl_2$	−1·00
Al + 1Zn	−0·96
Al + 7Mg	−0·89
Al + 5Mg	−0·88
Al + 3Mg	−0·87
$MnAl_6$	−0·85
99·95 Al	−0·85
Al + 1Mg_2Si	−0·83
Al + 1Si	−0·81
Al + 2Cu	−0·75
$CuAl_2$	−0·73
Al + 4Cu	−0·69
$FeAl_3$	−0·56
$NiAl_3$	−0·52
Si	−0·26
Cu	−0·20

treatment of the high-strength Al–Mg–Zn alloys, are more or less dangerous depending on their distribution through the alloy.

Thus, atmospheric grain boundary corrosion of the very common malleable and casting alloys of aluminium, zinc and magnesium is a phenomenon which may be explained relatively easily in terms of differences in electrochemical potential between solid solutions of precipitated intermetallic phases and the adjacent deficient zones. So-called layer corrosion may be explained similarly. This is a special case of grain degradation produced by corrosion, and is related to the lamellar distribution of individual alloy phases induced during rolling [6].

The most important methods of protection against such corrosion phenomena are obviously suitable heat treatment of the alloy and the choice of such manufacturing processes as will lessen the chance of producing undesirable changes in structure, e.g. by welding or aging processes. Another process method is the application of protective layers. The best example of this is pure aluminium plating of tempered aluminium alloys, which are susceptible to grain boundary corrosion and are an indispensable material for aircraft construction.

4.3 Stress corrosion cracking

Mechanical stressing of metals and alloys, especially in the region of plastic deformation, has a known influence on their structural properties and the distribution of composition in them. Stress corrosion is a phenomenon in which extremely localized cracks form very rapidly due to specific action by electrolyte components and specific super-critical tensile stresses (the internal stress of the material is often sufficient for this). The cracks may be either transgranular or along the grain boundaries. In the space of this chapter, it is not possible to examine the numerous stress corrosion cracking theories, which take into account such influences as e.g. energy of structure defects, extreme acceleration of plastically-flowing metal corrosion, because of the high mixing density, and tunnel formation into segregated alloy impurities at the surface. From the point of view of the electrolyte, specific adsorption at high-energy points, complex formation, disruption of the passive layer by dissolution or deformation (e.g. due to hydrogen evolution), etc., are deemed to be of importance.

In spite of the multiplicity of these factors, the current state of the knowledge of this extremely complicated form of corrosion can be summarized into specific categories [7]. No generally-valid mechanism for this phenomenon has yet been found; each stress corrosion-susceptible alloy-electrolyte system must be dealt with individually in terms of the factors involved.

Two basic mechanistic paths can be differentiated. In the first, the anodic dissolution process plays a governing role, and cathodic polarization can limit or prevent the cracking. In the second, hydrogen evolution and the concomitant deformation is especially important, and cathodic polarization accelerates the cracking.

Crack formation caused by adsorption is related to stress corrosion cracking, but the two are probably not wholly identical [8]. Stress corrosion cracking of ionic crystals e.g. silver chloride in solutions containing $AgCl_4^{3-}$ [9], or of organic polymers, such as poly-olefins, in surface-active environments, show that electrochemical processes do not always play the decisive role. Since adsorption of ions and electro-neutral solution components is potential-

(e.g. in adsorption of stimulating anions), it was investigated whether primary adsorption was a necessary first step of the process, and whether there was a close relationship between stress corrosion cracking and adsorption-stress crack formation [8]. This was hardly successful, however; without doubt, metallurgical factors play a more important role in many cases.

In stress corrosion cracking, the induction (up to crack initiation) and crack formation periods can be differentiated. Different mechanisms have been proposed and discussed for the beginning of crack formation [8, 9]:

(a) Reduction of the surface energy by adsorption. The relevant sorption energies must be greater than about 100 kcal mol^{-1}, which seems hardly likely. If surface compounds are formed by chemisorption (which these energies would require), then it is difficult to understand why extremely pure metals are not susceptible to stress corrosion cracking, and why there is specificity of electrolytes against particular groups of alloys.

(b) Localized disruption of the kinetic stability of the surface layer can perhaps be responsible for crack initiation. Into this theory can be further included e.g. local disruption of the surface film by pit initiation, preferential dissolution of the layer at segregation points, accumulation of heterogeneities, brittle fracture of the surface passive layer by mechanical stress (e.g. from hydrogen embrittlement), etc. All these mechanisms assume that there is local disruption of a passive state on the surface. Each local depassivation must be preceded by primary adsorption of activating solution components. The stress state can perhaps affect this process, as probably can the mechanical (elastic and plastic) and electrochemical (isoelectric point, ion exchange capability) properties of the passive layer, which are little-known to date.

Development of the stress corrosion crack proceeds via a complex simultaneous action of chemical processes involving the metal in the initiated crack, specific components in the electrolyte, and physical processes in the metal. The different possible explanations of the rapid crack formation process can be summarized as follows [9]:

(a) Intensification of tensile stress by the formation of solid corrosion products in the crack. While this effect is hardly decisive, it cannot be ignored completely.

(b) The different mechanisms of plastic flow are involved. When relatively few slip planes with high inhomogeneity density are present, a different stress corrosion cracking behaviour may be expected compared with when the flow is by a more symmetrical movement of the inhomogeneities. Also related to this theory is the hypothesis that there is a connection between structure defect energies and transgranular stress corrosion cracking. It is expected that alloys with lower structure defect energy are more susceptible to stress corrosion cracking than are those with a higher energy. Lower structure defect energy is connected with a hindering of transverse shear in plastic flow [9], so that isolated slip planes with extremely high inhomogeneity density arise. These can serve as preferred dissolution sites.

(c) Accelerated dissolution due to stress energy (in the elastic region). This effect appears more applicable to an explanation of stress corrosion cracking of non-crystalline materials such as glass. The small increase in the corrosion of metals in the elastic stress region is obviously insufficient to be responsible for the extremely rapid crack formation.

(d) Acceleration of the process by plastic flow. Since such flow causes a high inhomogeneity density, due to newly-forming slip surfaces, a marked acceleration of the dissolution process is understandable, and can be demonstrated experimentally [10].

(e) Selective corrosion acceleration at points of phase segregation. This process, which could also be designated as stress-promoted grain boundary corrosion, is often observed on precipitation-hardened aluminium alloys. It can also be observed on single-phase alloys, however; for example, on pure iron in which interstitial carbon and nitrogen compounds have formed by interaction with the environment. When such a deposit lies partly in the surface plane, there is very rapid tunnel-like corrosion which leads to crack formation.

Numerous other effects and mechanisms could be discussed here. The possibility of accelerated hydrogen absorption in the crack tip, with consequent embrittlement, selective dissolution of the base metal, and electrochemical deposition of noble alloy components on the crack walls, which should lead to protection of the walls and accelerated dissolution of the crack tip, all appear to be important theories worth noting. Figure 42 is a schematic representation of the different theories [7].

Even if atmospheric corrosion conditions are not particularly aggressive, the danger of stress corrosion cracking cannot be dismissed, and this is important in limiting the acceptable uses of high-strength alloys e.g. in aircraft construction. Different types of materials are especially endangered by stress corrosion cracking in salt-containing marine and coastal atmospheres. It is important that the possibility of stress corrosion cracking of high-strength steels and aluminium alloys, in particular, should be recognized. Zinc and magnesium alloys can also be attacked by this form of corrosion under certain conditions. On the other hand, stress corrosion cracking of austenitic Cr–Ni–steels, which is found often in aggressive liquids, is very rare in the atmosphere.

The remainder of this chapter will be devoted to a more comprehensive discussion of the most important types of material, especially high-strength steels, hardened aluminium alloys, and brass.

4.3.1 Stress corrosion cracking of high-strength steels

In general, all steel types strengthened by heat treatments with elastic yield

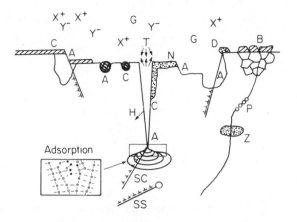

Figure 42. A schematic diagram showing the main points of theories of stress corrosion cracking [7]. X^+. cation; Y^-. anion; G. dissolved gas molecule; A. anodic surface or phase; B. breakthrough of passive layer; C. cathodic surface or phase; D. noble precipitate on surface; H. penetration by hydrogen; N. area made noble by selective corrosion; P. grain boundary precipitation; SC. stress–crack interaction; SS. stress–structure interaction; T. transport processes; Z. depleted zone. Inset shows the adsorption processes at the crack tip.

point values up to $140 \, \text{kp mm}^{-2}$ tend to corrode by stress cracking under atmospheric conditions [11]. The different steel types behave similarly, with no effects from variations in composition or structure. The elastic yield limit seems the most important factor for this phenomenon.

Aged martensitic steels always have a longer time to rupture than do other high-strength steels, including tempered stainless types. Austenitic steels, strengthened by cold-working, are very resistant to stress corrosion cracking in marine atmospheres.

It is very difficult to summarize the different mechanisms of stress corrosion cracking of high-strength steels. Both of the mechanisms discussed earlier occur in individual cases. Even the same steel may split because of hydrogen embrittlement or because of accelerated anodic dissolution at an initiated crack tip, depending on the heat treatment history. Stress cracks may be inter- or trans-granular in different cases. The effects of specific microstructural phases and their distribution have not been explained unequivocally as yet. In martensitic chromium steels, the distribution of chromium carbide and the associated chromium deficiency of the adjacent areas perhaps plays a role, and associated precipitates of martensite at the grain boundaries may induce stress corrosion cracking susceptibility.

The stress strength cannot generally be quoted as a limit value. Laboratory tests have yielded some limit values for certain steel types with defined heat pre-treatment, but these are significantly less than (mostly 50 to 75% of) the elastic limit, which implies a considerable restriction on the technical use of these steels.

4.3.2 Stress corrosion cracking of tempered aluminium alloys

In contrast to the complicated position of theories of stress corrosion cracking of steels, many more valid comments may be made on this form of corrosion of aluminium alloys. The cracks are always intergranular, and the mechanism is clearly electrochemical. Transition from grain boundary corrosion, as was discussed earlier, to stress corrosion cracking is gradual, and both phenomena may be mistaken for one another under certain conditions, the more so in that similar electrochemical differences due to particular structural phases occur in both cases. In contrast to grain boundary corrosion, which may possibly be present at the same time, and which develops mainly because of marked differences in potential between the different phases, stress corrosion cracking may be produced simply by higher crystal lattice disruption density or chemical composition gradients at the grain boundary (i.e. without phase formation). Aluminium alloys which are inclined to corrode at grain boundaries are thus less susceptible to stress corrosion cracking. There is no practical benefit to be gained from this, however, since the generally non-localized grain boundary corrosion still proceeds too rapidly in many cases to allow technical use of the alloy. When optimum heat treatment and other processes (e.g. working) are investigated to find the conditions giving least susceptibility to stress corrosion cracking, a reduction in general resistance to grain boundary corrosion and pitting must usually be accepted.

4.3.3 Stress corrosion cracking of copper–zinc alloys

Though the problem of 'season cracking' has lost its technical importance, a short review of this once widespread form of corrosion of cold-formed brass products should be given. Internal stress and the presence of a surface electrolyte containing NH_4^+ ions and dissolved oxygen is the decisive combination of factors for atmospheric stress corrosion cracking of work pieces with a mainly α-brass structure [12]. Ammonium ions play an especially important role on α-brass, whereas α,β-brass and β-brass may be attacked by stress corrosion cracking even without their action. The mechanism of the involvement of the ammonium ions in this corrosion seems clear. In their presence, and with co-action from dissolved oxygen in the electrolyte, copper reacts by [12]:

$$Cu + \tfrac{1}{2}O_2 \rightarrow CuO(ads) \tag{4-1}$$

$$CuO(ads) + 4NH_3 + 2H^+ \rightarrow [Cu(NH_3)_4]^{2+} \tag{4-2}$$

The complex ion $[Cu(NH_3)_4]^{2+}$ then dissolves further copper (or copper oxide):

$$[Cu(NH_3)_4]^{2+} + Cu \rightarrow 2[Cu(NH_3)_2]^+ \tag{4-3}$$

and the product from this reaction is reconverted to $[Cu(NH_3)_4]^{2+}$ by atmospheric oxygen and more ammonia. The process is thus autocatalytic. The dissolution of zinc is also caused initially by the action of the $[Cu(NH_3)_4]^{2+}$ ions.

If mainly α-brass is present, two variations of the stress corrosion cracking may be observed. The first occurs in alkaline electrolytes which have sufficiently high ammonium ion concentrations to prevent oxide films forming at the surface. The stress cracks in such cases are mainly intergranular on those alloys containing less than 10% zinc. On alloys containing more than 20% zinc, the cracks are mostly transgranular [13].

At lower values of pH (less than 8), surface oxide layers are more stable on thermodynamic grounds, and stress corrosion cracking occurs by their formation and local destruction. The cracks usually follow the grain boundaries for the whole range of compositions in α-brass.

The latter form is the one typically found for stress cracking of α-brass in the atmosphere. The ammonia content of the atmosphere is rarely so high that in the simultaneous presence of acid-forming atmospheric components (carbon dioxide, sulphur dioxide), surface electrolytes with a sufficiently high ammonia content and pH >7 could be formed, as is required for oxide-free stress corrosion cracking.

Brass types with high zinc contents and other alloying additions which have α,β- or β-structures tend to corrode by stress cracking not only in the presence of ammonium ion-containing electrolytes but also in pure water or sodium chloride solutions. The current view is that this cracking is caused by adsorption [12], since in absence of electrolyte these materials fracture transgranularly during creep testing, while introduction of electrolyte produces a far shorter resistance period and intergranular fracture.

Suitable heat treatment allows virtual removal of internal stresses from cold-worked brass products without impairment of their mechanical properties. Thus, the problems caused by an atmospheric stress corrosion cracking of brassware are now understood and have been made almost negligible.

5
Principles of Protection Against Atmospheric Corrosion

5.1 Derivation of protection methods

The knowledge of physical and chemical principles of atmospheric corrosion which has been gathered must be applied in deriving protection methods if they are to be the most effective possible.

Atmospheric corrosion is the result of predominantly chemical effects of reactive components of the atmosphere on the metal. This causes changes which impair the functioning (or appearance) of products made from the metal.

Proceeding from the discussions in Chapters 3 and 4 on hypotheses on mechanisms and kinetics of corrosion reactions, and considering the differing properties of the corroding metals and the atmosphere, principles of corrosion protection may be sketched out using a model based on the concept of a 'dominant influence' (or a dominant combination of influences).

It follows from the review in Chapter 2.3 of corrosion-affecting influences that the dominant factors embrace not only the material which is corroding but also the corrosive environment i.e. the atmosphere. There are special effects on the course of the corrosion due to the frequency and duration of periods of presence of surface electrolyte. The time of wetting may be deduced directly from meteorological data, and is dependent on the mass of the metallic object and its design. (This will be examined more closely in Chapter 7.) The course of the corrosion is affected further by the type and amount of atmospheric pollution by gaseous, liquid and solid species, given the assumption of formation of a surface electrolyte. Other atmospheric effects, such as temperature, air pressure, the absolute quantity of precipitation, and radiation, are of undetermined significance for atmospheric corrosion of metallic surfaces. The chemical composition, structure and surface character of the corroding metal should be known. Mechanical stress of the metal may play a supplementary role in development of dominant factor combinations.

Corrosion protection can be achieved by changing the state of the metal (especially at its surface) on the one hand or by changing the dominant effects of the atmosphere on the other. Since there are various technical methods for

85

setting up microclimates in the environment close to the metal surface (especially in closed chambers), it is possible to remove the influence of some dominant factor combinations e.g. presence of corrosion-stimulating species and 'super-critical' humidity, or condensation of atmospheric water vapour.

The nature of the metal or its surface may be altered in numerous ways to achieve improved corrosion resistance. Easiest of all, the most suitable of the many alloys whose corrosion behaviour under different atmospheric conditions are known can be chosen for the specific application. (Technical and economic criteria for this choice are discussed further in Chapter 7.) Only unusually are special alloys produced to have improved corrosion resistance to specific atmospheres and used technically to any great extent; an example is the so-called weathering steels.

If mechanical stresses (either static or dynamic) are involved in the dominant factor combination, especially if stress corrosion cracking is to be expected, the choice of material and its heat treatment and the design and method of manufacture of the product assume special importance.

The different types of coating of insufficiently resistant materials with other materials to produce more favorable properties can also contribute to protection of the metal against corrosion in different atmospheres. (See also Chapter 5.4.) The most widely used coatings are organic materials, chiefly paint and plastic layers. Oil, grease and wax, which are widely used for temporary protection, also belong to this group.

If the mechanisms of action of the different methods of protection against atmospheric corrosion are examined to classify them under the ideas discussed in Chapter 3 on the chemical basis of atmospheric corrosion, the following corrosion protection principles are found:

(a) Maintenance of the inherent protective oxide layer by protection of it from disrupting agents (water, stimulator species). Destimulation of the atmosphere, and some aspects of protective action by organic coatings (including those used for temporary protection) and use of inhibitors are based on this principle.

(b) Use of the inhibiting effects of the corrosion products formed. The action of zinc and cadmium coatings, among others, is based on this; it is also responsible for the higher corrosion resistance of rust-bearing steels.

(c) Use of impressed cathodic protection, though this idea is often over-emphasized. Because of the limited conductivity of the surface electrolyte, only insignificant impressed protection can be expected [1]. This protection mechanism can only operate over short distances and at large differences from the stationary corrosion potential (e.g. from zinc coatings on steel or other noble metals). In addition to the low ionic conductivity of the surface electrolyte, there is a limitation imposed on cathodic protection by time factors. The corrosion products formed change the potential difference between the 'sacrificed' metal (e.g. zinc) and the protected metal quite quickly, so that the electrochemical action decreases.

5.2 Destimulation of the atmosphere

It is apparent from Chapter 3 that atmospheric corrosion is caused by three components of the environment (i.e. atmospheric oxygen, water (liquid or vapour), and stimulator species) acting simultaneously, and in certain minimum amounts. In many cases, it is only necessary to remove one of these three dominant factors from the atmosphere, or to maintain it below the minimum required for its action, to exclude the danger of corrosion. Oxygen poses most problems here; its content in the atmosphere is constant, and there are considerable difficulties in devising a method for its removal from the microclimate near the surface which is to be protected. Only exceptionally is it economical to use a closed chamber with the air replaced by an inert gas to protect a metal. (This occurs mostly in long-term preservation of particularly delicate or valuable objects e.g. in museum conditions.) The removal of the oxygen must be complete, since 3% oxygen in the gaseous environment is sufficient to produce corrosion [2].

Partial removal of the water, as the second of the atmospheric components necessary for corrosion, is far simpler, cheaper, and consequently more widespread. Considerations of what level of removal of this atmospheric component is suitable must examine the following points:

The concept of 'critical humidity' (see Chapter 3) is very important here, though not sufficient in itself. This concept was discussed more closely in earlier sections. The critical humidity is not a value fixed for every case; it is basically set by the actual conditions for transformation of water vapour in the atmosphere into liquid water on the metal surface. As has been discussed already, its value at constant temperature can be reduced markedly if there are hygroscopic and/or soluble species present on the surface. In such a case, the value usually lies close to the saturation vapour pressure of a saturated solution of the species in question. Whether this species which reduces the critical humidity is deposited on the metal surface as a solid or as a dissolved precipitate from the atmosphere or as the reaction product from a gaseous pollutant is of little significance here, since treatment of the atmosphere to reduce its water content can only provide protection against corrosion in enclosed spaces, such as warehouses, workshops, or packages. Another source of condensation promoters is the manufacturing process. In working with metals, many chemical species are used whose residues can cause pronounced condensation to leave a layer of liquid water on the metal surface. Pickling or degreasing solutions, cooling emulsion residues, and phosphatizing or galvanizing baths all contain species which are water-soluble and hygroscopic.

Another frequent cause of localized electrolyte layer formation, which is sufficient to sustain corrosion, are traces of sweat which have been deposited during handling of the metal sample. The major components of sweat, which are sodium chloride and organic acids, promote condensation and stimulate the corrosion process by their chemical action.

Physical water layer formation at the surface by adsorption onto the colloidal fraction of the corrosion products close to the product-metal interface is of little interest in this study. In practical terms, air drying is only useful for protection against formation of small amounts of corrosion products, and in particular for temporary protection or for protection of corrosion-prone objects which can be otherwise protected only with difficulty.

In arriving at a technically useful value of critical humidity, allowance should always be made for the possible or probable presence of water-soluble hygroscopic species on the metal surface.

Up to this point, the discussion has concerned those effects which reduce the value of critical humidity at constant temperature. Frequently, smaller or larger short-term temperature variations must be considered. These can have pronounced effects on the relative humidity value at which surface electrolyte forms by altering the long-term stability of the absolute humidity. It is clear from Table 3 that if the atmospheric humidity is high, even a small drop in temperature can lead to formation of liquid water because the higher relative humidity which now exists exceeds the dew point, or at least exceeds the critical humidity.

This danger is far greater in the tropical warm moist regions than in other regions, and greater in the warm season than in the cold. The physical reason for this is the exponential increase in water vapour pressure as temperature increases. At the higher temperatures, at the same relative humidity, the absolute water content is always higher, and a quite small decrease in temperature causes a significant rise in relative humidity, perhaps to reach the critical humidity for corrosion or even the dew-point.

This is especially important in setting permissible humidity limits in e.g. parcels with respect to danger of corrosion. Very wide temperature ranges must be anticipated for transport lasting for long periods (e.g. overseas surface mail), and the maximum water content of the enclosed environment in such cases must be so restricted that the dew-point cannot be even approximately attained.

In discussing practical methods of air treatment to afford protection against corrosion by limiting the water content, two cases should be differentiated:

1. Reduction of atmospheric water content in incompletely closed systems, such as a warehouse or workshop. In such cases, a relatively stable temperature may be assumed, which will be regulated by heating in winter. Critical humidity determination is not difficult in such cases. Simple measures may be applied to exclude species which promote electrolyte formation and hence the probability of corrosion, and to maintain the humidity value just below the critical humidity. Experience suggests that keeping the humidity below 80% (and at lowest 60 to 70%) is sufficient. The climate of the region determines what measures should be necessary; it is sometimes possible to circumvent the use of costly equipment simply by installing a well-chosen air-conditioning system.

2. Special problems arise in warehouses with high throughput, particularly in the colder seasons, due to the danger of corrosion when the critical humidity

or dew-point is exceeded frequently. Frequent opening of doors leads to intro-duction of cooler air, which presents no great danger in itself, but may cool metallic objects near the door onto which condensation may then occur. Under-cooling of objects in this manner, leading to a localized increase in relative humidity, is often a cause of unexpected corrosion. Still more dangerous is the introduction of cool metal objects into an already warm room which is not completely dry. Especially if the object is of large mass and so high heat capacity, its temperature hysteresis often produces a long-lived under-cooling, which in turn causes condensation of water onto the surface. 'Sweating cor-rosion' e.g. of zinc, can often be attributed to this; for example, galvanized steel roofing sheets are often attacked very intensively on the inside, since damp, warm air under the roof loses water by condensation onto the surface which is cooled by the external colder air.

Destimulation of the atmosphere thus cannot be done by rule of thumb. Each time this very elegant and effective corrosion protection measure is applied, extreme values of possible temperature or humidity variations, heat capacities of the objects to be protected, and other considerations like these must be calculated and allowed for.

Exclusion of gaseous corrosion-promoting species (stimulators) from the metal surface is difficult to apply as a corrosion protection method. As was explained more fully in Chapter 3, very small amounts of such stimulators are sufficient to produce corrosion and accelerate the process. Their origin is not only atmospheric (i.e. sulphur compounds, salt particles, etc.); operations on the metal and handling of the finished product also lead to contamination of apparently pure surfaces by corrosive species, which is difficult to control or remove. Only extremely careful control of all the stages of the cleaning opera-tions and cleanliness in the handling of the finished goods (e.g. with frequently-washed cotton gloves) leads to good results. This is especially true for delicate, complicated instruments and for mechanisms for precision instruments, optics, low-current devices, etc., but these techniques pay dividends wherever corrosion-sensitive metals (e.g. carbon- and low allow-steels, or zinc or magnesium alloys) are to be stored, transported, or required to remain free from corrosion while in the factory, with no permanent surface protection.

Shielding of metallic goods from gaseous stimulators is used only rarely, and mostly in packaging. Packing materials (paper) impregnated with copper-chlorophyllin derivatives are used to exclude hydrogen sulphide, which causes formation of invisible sulphide films on copper or silver, even at very low con-centrations. Chlorophyllin derivatives react with very small amounts of H_2S to give stable compounds, so that the penetration of this gas from the external atmosphere is temporarily hindered.

5.3 Corrosion inhibitors for protection against atmospheric corrosion

As has been discussed already, the chief principle of the use of inhibitors for protection against atmospheric corrosion is the maintenance or strengthening of the inherent, though usually rather limited, protective oxide layer which forms very rapidly on all base metals on contact with the atmosphere. The use of inhibitors is thus limited to 'technically pure' surfaces, which are to be protected either only briefly or for longer periods against formation of corrosion product layers which impair their use.

Since the surface electrolyte formed by condensation of atmospheric water vapour normally has an approximately neutral pH, the choice of suitable inhibiting species is rather limited. It is well-known that the problem of finding effective organic corrosion inhibitors for neutral solutions has still not been satisfactorily resolved.

The choice is thus limited to using those groups of species which have been found to reduce the aggressive effects of neutral solutions, or to modifications of these. It is not important for this discussion whether they are used as contact inhibitors or vapour phase inhibitors for protection against atmospheric corrosion; the only difference between the two types is that a contact inhibitor must be used in direct contact with the surface to be protected, while a vapour phase inhibitor can always reach the surface by evaporating and condensing onto it.

The surfaces most sensitive to corrosion are undoubtedly those on iron (or steel or cast iron). General data on passivation of iron shows that inherent passivity may be attained by raising the pH value of the electrolyte or by increasing the stationary corrosion potential (Figure 12). The commonest inhibitors of steel corrosion are chosen on this basis; they are either species which act by oxidizing other species or they hydrolyse to give weakly alkaline solutions. Often, the two principles are combined. Amongst the most important of these inhibitors are nitrite and chromate ions, and more rarely vanadate, tungstate, and molybdate ions (all of these acting by oxidation). Examples of inhibitors which act by hydrolysis are sodium carbonate and phosphate.

The mechanisms of action of these inhibitors are not, however, as simple as might be anticipated from Figure 12. It is certain that oxidizing inhibitors promote passivity, but they must also be relatively easily reducible i.e. the exchange current density of the electrode reaction corresponding to their reduction must be of a suitable magnitude. This is the reason why permanganate ions do not produce passivity. Cartledge, in particular, has shown that they must be easily adsorbed [3]. The oxidizing inhibitor definitely participates in the formation of the passive layer during passivation, since it forms a considerable fraction of this layer in its original or reduced form [4]. This phenomenon is obviously related to potential-dependent chemisorption of anions during the passive layer formation processes [5].

The second group of inhibitors are those which come into contact with the metal mainly by adsorption or chemisorption or by formation of water-resistant compounds. Phosphate, benzoate, and most inhibitors which are used in preservative oils, fats and waxes act in this way. Benzoate and phosphate are especially interesting; their action is affected little by variable presence of oxygen, and they exert their inhibitive action by forming very thin protective layers of ferric complexes with these anions [4].

As was indicated earlier, the sole difference between the actions of contact and vapour phase inhibitors is that contact inhibitors must be used in direct contact with the surface, while vapour phase inhibitors act after adsorption from the atmosphere. The most important reasons for their inhibition remain the same, however. Active components of the vapour phase inhibitors involve oxidizing, easily reducible anions (especially NO_2^-) or groups which act by adsorption, such as benzoate ions. To obtain vaporizability of these inhibiting groups, they must be included into compounds which have relatively high vapour pressures at normal environmental temperatures. This is found only rarely in simple salts; one example is ammonium benzoate, which is quite effective under certain conditions. The volatility of ammonium nitrite can also be used, though preferably indirectly since this salt is not stable (and often decomposes quite briskly). It is therefore formed gradually by reaction between amine-containing species and sodium nitrite; mixtures of sodium nitrite and urea or hexamethylene tetramine are some examples.

Ferrous metals are generally well-protected when the vapour phase inhibitor contains the passivating NO_2^- group and amine groups. Nitrites of cyclic amines have proven to be particularly effective inhibitors of atmospheric corrosion of ferrous metals. Compounds with desired volatilities can be produced by combining nitrite ions with different amines, so that their evaporation rates and 'supply' can be regulated.

Dicyclohexylamine nitrite has proven to be the best inhibitor of this type:

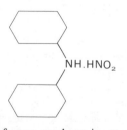

NH.HNO$_2$

It is used widely in different forms, such as in an alcoholic solution spray or as an impregnant in packing paper, to give protection in small spaces. The organic part of the inhibitor produces the required controlled volatility, and produces enhanced protection as well. Cyclohexylamine is probably a relatively more effective inhibitor of steel corrosion, in terms of adsorption and alkaline hydrolysis.

Other compounds of cyclohexylamine besides dicyclohexylamine nitrite are also used, though more rarely and normally as admixtures. The higher evaporation rate of cyclohexylamine carbonate can be exploited if brief protection is required and large distances between the inhibitor source (the packing material) and the object to be protected are to be overcome. This inhibitor is sometimes admixed to dicyclohexylamine nitrite to ensure continued protection after sealing-off of the microclimate. Good results are also achieved with ethanolamine carbonate or ethanolamine benzoate (in mixtures with ethanol). The most important vapour phase inhibitor of ferrous metal corrosion is still undoubtedly the so far unexcelled dicyclohexylamine nitrite, however.

This inhibitor is not suitable for simultaneous protection of other metals; in fact, it may promote their corrosion (e.g. of zinc). In such cases adsorptive inhibitors such as ethanolamine benzoate are to be preferred, even though they are not as effective as is dicyclohexylamine nitrite in protecting iron (because of the suppression of the passivation effect).

Experiments to develop universally effective volatile inhibitors of atmospheric corrosion in which the chromate anion is converted into organic, easily-vaporized compounds have had no success to date [6]. The compounds are not very stable and hydrolyse easily, and their oxidative ability is so high that they cause destruction of the packing material (e.g. considerable embrittlement of paper).

5.4 Alloying to increase atmospheric corrosion resistance

The atmosphere is not, in itself, as aggressive an environment as are such media as acids, salt solutions or sea-water. This is the main reason why special alloys with improved atmospheric corrosion resistance have rarely been developed. The reverse is true of the more aggressive media, where special corrosion-resistant alloys, with composition and structure adapted for the specific corrosion expected, are often used. Corrosion-resistant alloys which can be used in their passive states under these more aggressive conditions normally remain sufficiently resistant to atmospheric conditions. This applies particularly to high-alloy steels with suitable chromium contents, and the large class of stainless chromium-nickel special steels.

The 'weathering' steels represent the single important case where alloying additions are made consciously with the aim of increasing the resistance to atmospheric corrosion. Though this type of steel has been known for many years, it has only recently come into widespread use. The chemical composition of these steels was not developed from a systematic research programme, but came from accidental use of raw materials with high contents of the favorable alloying elements.

The weathering steels cannot be compared in any way with the high-alloy permanently passive steels. The low-alloy steels rust when exposed to atmospheric conditions, though significantly more slowly than does normal carbon steel. Some statements can be made on the actions of the individual and combined alloying additions, though they must be based to date mainly on statistical evaluations. The chemical basis of the protection effect is only little understood. However, the various ideas on the protection mechanisms must be examined.

The active additions are mainly those of copper, nickel, chromium, silicon, and phosphorus, though beryllium and others may be used.

The favorable action of copper has been known longest [8] (Figure 43). The so-called copper steels have been recommended for a long time for bridge-building and similar applications. It appears, though, that there is a definite lower concentration limit (about 0.12%) to the favourable influence of copper. This is apparently related to the deleterious effects of sulphur [8, 9]. The

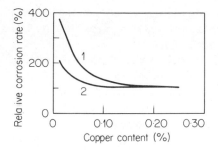

Figure 43. Effects of copper contents of steels on atmospheric corrosion rates of the steels.
1. high sulphur content; 2. low sulphur content.

unstable iron or manganese sulphide inclusions can possibly form nuclei for sulphate accumulation, since oxidation to iron sulphate occurs easily. Copper sulphide is more stable, however. If its slower oxidation does nevertheless occur, the reaction product is most probably sparingly soluble basic copper sulphate, which then undergoes no further hydrolysis. It thus seems that copper lessens the injurious effects of sulphur on steel. This theory can be confirmed by radio-chemical observation of sulphate nest formation in the laboratory. If steel plates are inoculated with iron sulphate in soluble form and then allowed to rust in a pure damp atmosphere, sulphate nests form more or less densely depending on the amount of sulphate. On low-alloy steels, this phenomenon is much less apparent [10].

Copper can also reduce the activity of sulphate or chloride anions, which strongly promote corrosion. As was explained more fully in Chapter 3, the mechanisms of action of these stimulators on iron are unique to iron and 'auto-catalytic'. During rust formation, significant amounts of the stimulator are reliberated. If, however, there are certain concentrations of cupric ions present at the steel-rust interface, which is the crucial one in this system, there is the possibility of reducing the activity of the aggressive ions by binding them into basic salts. This reduces the rusting rate, since it is a function of the activity of these ions [11]. The favourable action of the copper ions can also be explained in other ways. Colloidal hydroxides are formed at the steel-rust interface as solid products in addition to the usual normal and basic salts, and these hydrox-ides convert only slowly into the crystalline final form of the rust (α- and γ-FeOOH), which is no longer active. The colloidal form represents a source of water and sulphate at the reaction interface, since both the water and the anions are bound only by relatively weak adsorption forces and are thus relatively easily available for the corrosion process. The stability of the colloidal layer is obviously related to its anion content (since it is formed by simultaneous oxidation and hydrolysis of the normal salt). If the anion content is reduced by binding of anions into basic copper salts, the colloidal layer transforms into inactive rust and the water evaporates more easily, so that the total time of surface electrolyte presence is shorter. This theory is supported to a certain extent by experience: the favourable effect of copper is found to be stronger in industrial atmospheres than in pure ones (suggesting a connection with the stimulating anions), and the best results are achieved if the rust on the low alloy- or copper-steel is able to dry out periodically.

The mechanisms of protection by the other elements which the statistical study suggests to have beneficial effects on the corrosion resistance of low alloy steels are not well understood. Nickel can form sparingly soluble basic sulphates like those on copper, and nickel sulphide inclusions in steel are more stable than are iron or manganese sulphides. The favourable influence of phosphorus (which appears only if copper is also present) is difficult to explain. It may be due to gradual formation of insoluble phosphates, which makes the active rust layer less porous and thus crystalline. Silicon may have similar beneficial effects due to formation of silicic acid (or perhaps a complex phosphoric-silicic acid compound). The action of low chromium contents during formation of protective rust layers is unexpected. Low chromium contents in steels raise the corrosion rates of steels in their active regions; however, Pourbaix and others have shown that atmospheric corrosion of iron occurs mostly in the passive potential region, with the passivity disrupted only locally by activating anions [12]. The chromium possibly aids formation of resistant primary layers, so that the active corrosion surface (where there is primary formation of normal salts) is kept to a minimum. The classical combination of alloying additions (copper, nickel, chromium, phosphorus and silicon) in this type of steel is undoubtedly the best, even though the mode of action of the individual and combined alloying elements is in no way clarified as yet.

For completeness, two further theories of the protective action of copper should be discussed. Tomashov [13] suggests that the copper acts as a cathodically active inclusion by promoting the cathodic part-processes. This would allow increased anodic partial currents, and so promote passivity. This hypothesis is based on experiments on steels to which noble metals have been added (e.g. with added platinum or palladium), which indicated a protective effect similar to that from platinum- or palladium-alloyed stainless 18:9–Cr–Ni-steels or titanium in acid. The second theory maintains, without explaining why, that rust-covered copper-containing steels have a higher critical humidity for corrosion than does normal building steel, so that the total duration of corrosion periods is shorter [14]. This is in accord with the ideas discussed earlier on the binding of sulphate ions into insoluble salts i.e. limitation of the amount of hygroscopic normal salts or water-absorbing colloidal rust layers.

Some results from U.K. suggest that copper-free low-alloy slow-rusting steels can be made. The alloying components e.g. Be–Al or Cr–Al, are not economically or metallurgically profitable, however. Little can be reported to date on the mechanisms of the protective effects of such alloying additions [15].

Other metals besides iron have been protected against atmospheric corrosion by alloying. Examples of these include additions of zinc, cadmium or beryllium to silver to improve resistance to sulphide formation, or raising of the corrosion resistance of electrolytically deposited zinc coatings by enrichment with co-deposited titanium. The mechanism of this latter effect on zinc remains to be clarified.

5.5 Protection by metallic coatings

The idea of laying a corrosion-resistant metal over a less resistant one to improve the resistance of the latter is the basis of one of the oldest methods of protection against corrosion. The multiplicity of technical methods for this process makes this type of protection particularly attractive, since sensible processes can be chosen for quite different metals depending on the size and number of products and on the aesthetic and functional properties required. The choice of the coating method is extremely wide; the following methods are available: dipping into the molten metal, thermal spraying, thermo-diffusion processes, electrolytic deposition from solution or molten salt, catalytic reduction in aqueous solution, vacuum deposition, powder metallurgy methods (e.g. by melting an applied layer of powder metal), plating, application of powdered metal in an organic or inorganic binder, etc.

The coating will possess properties which will depend not only on the type of metal used as coating but also on the coating technique. In particular, methods which involve high temperatures produce different properties because of reciprocal diffusion processes in the base metal-coating metal system. Coatings produced by thermal spraying invariably contain pores, since the coating is an accumulation of individual drops which have been deformed in the plastic state. The texture of coatings produced electrolytically is quite different from that on coatings produced by metallurgical techniques. Electrolytically and chemically deposited metal layers always contain traces of elements or organic additives which are present in the plating bath to aid the deposition process (e.g. phosphorus or boron in chemically deposited nickel, sulphur in bright nickel electrolytic coatings, etc.). These traces have marked effects on the corrosion behaviour of the coating in many cases.

Before the protective mechanisms of metallic coatings are examined more closely, the corrosion reactions of the whole base metal–coating–atmosphere (or –surface electrolyte) system must be analysed. When solid corrosion products from the coating metal are produced during the corrosion, the effects of these must also be considered. This analysis must also assume the (technically impossible) production of completely homogeneous coatings (i.e. coatings of uniform thickness and porosity) which have uniform properties independent of distance from the base metal. During corrosion, conditions in a real system will change. Removal of irregularly thick protective layers exposes the base metal (or a layer which is richer in the base metal). Solid corrosion products can block up inherent pores, or pores formed during corrosion, which expose the substrate. In multi-layer coatings, such as those produced electrolytically, the electrochemical relationships between the individual layers which make up the coating also play a very important role.

Corrosion in the substrate–layer–atmosphere system is basically an electrochemical process, just as is ordinary atmospheric corrosion (Chapter 3), but there are important cases where electrochemical and chemical reactions follow one another. Electrochemical principles of atmospheric corrosion cannot be ignored when explaining the protective effects of the metallic coating, even if the designation of the coating metal as anodic or cathodic with respect to the substrate is not clear-cut. As was mentioned earlier, the electrochemical properties of the substrate–coating–surface electrolyte system change during the corrosion, especially if solid corrosion products are formed. In no case can there be drawn definitive conclusions from the frequently-quoted electrochemical series of metals, whether they are the standard potentials or stationary potentials in e.g. sea-water or some other electrolyte. It is thus necessary to discuss each substrate–coating–surface electrolyte system individually. Iron (or carbon steel or cast iron) is the metal most used in technology, but at the same time it is the metal most susceptible to corrosion under atmospheric conditions (see Chapter 3). It is therefore desirable that metallic coatings on steel or cast iron should be given special attention.

Two basically different principles are used as criteria in the protection of iron against atmospheric corrosion. For some equipment, the main requirement is that the coating should maintain the mechanical properties (especially the different strength properties) required in construction. Examples here are construction steels or machine parts; in these cases, the surface state, and especially the appearance, plays a minor though not negligible role. In the second case, corrosion protection by the metallic coating is mainly to maintain the 'metallic' state of the surface. This might be required to enable the product to keep certain electrical or functional properties, or may be simply decorative. This latter use is often very important; for example, the fenders of a motor vehicle hardly lose their mechanical strength due to atmospheric corrosion during their expected 4 to 7 year 'life', so that coating of them is mainly as decoration.

Other technical requirements may indicate a reason for use of coatings. In machinery which must be dismantled periodically (e.g. for installation of replacement parts), there must be small enough atmospheric corrosion attack on the old parts to avoid the need to use special measures for their removal. This involves questions regarding the choice of the best protective measure for a particular application (which are dealt with in Chapter 7), but such criteria should also be mentioned here in the context of the mechanism of protective effects of metallic coatings on iron.

The most important coating metals for strength maintenance of iron are zinc, cadmium and aluminium, and more rarely nickel, copper, lead, tin, some alloys (e.g. Zn–Fe or Zn–Sn), or combinations of these. During long exposure periods in the atmosphere, solid corrosion products are formed on the surface of the protective metal coating, which alter its metallurgical character and its appearance. This does not, however, affect the maintenance of the strength of the underlying metal.

It is more difficult to select a coating system if the iron (or steel) must be protected against formation of all solid corrosion products. There are two possibilities here: either a metal with a highly stable passivity (e.g. chromium) or a noble metal (gold, platinum, etc.) must be chosen for the coating. These are not attacked by the atmosphere because of their thermodynamic stability. It is obvious that there must be complete freedom from pores here.

5.5.1 Zinc coatings on iron and steel

The protection of iron by zinc coatings is the most traditional and most wide-spread method. Despite the many unclear points in the explanation of the mechanism of its action, it is clear that the anodic character of the zinc with respect to the iron is rather over-emphasized. Continuing work [1] shows unequivocally that while the action of a zinc coating on steel as a sacrificial anode cannot be completely neglected, the long life of zinc-coated objects may be ascribed chiefly to the favourable corrosion products of the zinc coating.

The corrosion of zinc was dealt with comprehensively in Chapter 3. Basically, zinc corrodes in most atmospheres by a passive mechanism, which is the reason for the order-of-magnitude smaller corrosion rate of zinc compared with that of iron. The corroding zinc becomes covered with a layer of zinc oxide and hydroxide, which is destroyed gradually at the interface with the atmosphere (mainly by conversion into more soluble basic or normal salts which dissolve away or fall off). Because of this, the corrosion rate and thus the life of the zinc layer is stoichiometrically related to the quantity of corrosion stimulators taken up from the atmosphere [16]. In contrast to the autocatalytic renewal of the stimulator during corrosion of iron, whereby one molecule of sulphur dioxide or sodium chloride may remove up to 100 atoms of iron, the one stimulator molecule reacts with at most one zinc atom.

The duration of the protective action by zinc coatings can be divided into three periods [1]:

In the initial period, the actual protective layer of corrosion products gradually makes its presence felt. During this time, there is an electrolytically-based, very restricted, long-distance protection, and this protection may reach up to 1 or 2 mm in favourable conditions. This is how cut edges on thin hot-galvanized plates are protected from rusting, and how pores in thin electrolytic or metallized zinc coatings on steels are rendered ineffective. It can be deduced easily why the electrolytic long-range action is so limited; electrolyte layers which form from precipitation, condensation or adsorption are usually very thin i.e. the amount of water in the layer is small. It is further likely that the layer will have low conductivity, especially on zinc, since in atmospheres which are not very damp, zinc and its primary corrosion products have a very high binding force for formation of compounds with the anions present. Basic salts with a very low solubility product are formed rapidly.

Experiments by Evans [17] and Bartoň and Veselý [18] appear to suggest a further possible explanation of the electrochemical protective action of zinc with respect to iron. If stimulating anions are present in the electrolyte, they are transported to the anode, the zinc, by the current flow set up due to the zinc–iron potential difference, and harmless basic salts are formed. This principle is also used in practice [18]. The extremely stable and tenacious sulphate and chloride accumulations are very dangerous if the surface pretreatment of pre-rusted steel before painting of the surface does not include their removal. If the accumulations are permeated with zinc particles (e.g. by thermal spraying),

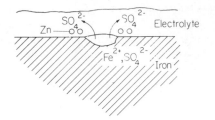

Figure 44. Schematic representation
of binding of sulphate ions by a zinc
coating on iron.

however, the soluble sulphate is soon converted into basic salts. The coating
seals itself, has good cohesion and adhesion, and can be painted without danger
of under-rusting, even when it is applied quite thinly. It may be assumed that
the long-distance protective effect of zinc coatings e.g. at the cut edges mentioned
above, is due partly to this action. The initial period ends when a coherent
layer of corrosion products has formed, which seals off inherent pores and takes
over the protective role.

Under certain conditions, however, no protective layer of products is
formed. During uninterrupted 'sweating' (condensation) with no periods of
drying out, soluble, easily removable 'white rust' forms, and very soon disrupts
the zinc coating. This corrosion phenomenon is often accelerated by the presence
of organic vapours. Čermáková [19] and other authors have described the
injurious effects of e.g. formic or acetic acids (the former also being formed by
catalytic oxidation of formaldehyde at the zinc surface. Formaldehyde often
occurs in enclosed spaces, e.g. when incompletely-condensed polymers or
glues containing formaldehyde are used).

The duration of the second and by far the most important period of the
'life' of zinc coatings is determined by the reactions between the electrolyte
and the already-formed corrosion products. As has been mentioned often
already, the process during this period may be described by the basic kinetics
of corrosion in the passive region, and depends on the amount of corrosion
stimulator reaching the corrosion product-atmosphere interface. It was shown
in Chapter 3 that primary adsorption processes (concurrent adsorption of
passivating and disrupting molecules and ions) can also play a role here. It
may be deduced from this that, just as for iron, there is a critical limiting concentration of stimulator for corrosion of zinc coatings, and exceeding of this
value leads to accelerated removal of the coating [1] (see Figure 40).

When the zinc coating has been removed so much during the second,
long-lasting, period that iron is exposed, at initially small and later larger
areas, the third and final period of protection by the coating begins. The
mechanism of this is similar to that during the first period. Long-range protection begins again, whether based on direct electrochemical action of the
remaining zinc as a sacrificial anode or transport of stimulator anions to the

zinc and formation of innocuous basic salts. Experience suggests that the chemical binding of the anions is still strong during this time; for instance, at the beginning of the third period, originally zinc-coated iron which has become partly rusted may be painted without danger of later under-rusting.

It follows from all this that the life of zinc-coated iron objects is directly proportional to the amount of zinc present (and thus to the thickness of the coating) and indirectly proportional to the aggressiveness of the atmospheric environment. The different methods of applying the coating are therefore of minor importance; coatings produced electrolytically or by metallizing or hot galvanizing which are of equal weight are of equal protective use, while the zinc-rich alloy layers formed during hot galvanizing have a similar protective ability to that of zinc.

5.5.2 Cadmium coatings on iron

The chemical nature of cadmium is very similar to that of zinc, especially in aspects regarding protection of steel in the atmosphere. Cadmium is a more noble metal than zinc, and so always affords less electrochemical protection to iron (i.e. with cadmium as anode) than does zinc. The formation of sparingly soluble oxides, hydroxides, and basic salts as corrosion products is essential for functioning of cadmium coatings. Atmospheric corrosion of cadmium is again by a passivation mechanism, with the rate-determining step occurring at the product-atmosphere interface. There is a stoichiometric ratio between the amount of stimulator arriving at the surface and the metal removal during corrosion. The end-product is basic cadmium sulphate in SO_2-containing atmospheres and basic cadmium chloride in atmospheres containing chloride. These are dissolved away or fall off as soluble or slightly-adherent products, so that the thickness of the oxide layer is maintained at the cost of the cadmium. Cadmium is a heavier metal than zinc, so that on a 'stoichiometric' basis, their atmospheric corrosion should lead to faster removal of a cadmium coating than of zinc. This is found in practice; especially in industrial atmospheres, in which sulphur dioxide is the dominant corrosion stimulator, cadmium coatings are shorter-lived than those of zinc of equal weight. In chloride-containing atmospheres, highly adherent, slightly soluble basic chlorides are formed on the surface of both zinc and cadmium, so that the expected higher rate of corrosion of cadmium coatings is not observed [20, 21].

The frequently-stated view that cadmium coatings are a superior alternative to zinc coatings under atmospheric conditions is not confirmed as generally valid by tests, so that the use of thinner cadmium coatings than of zinc coatings for use under the same conditions, which is often suggested in specifications, cannot be recommended.

In particular cases, cadmium coatings are indeed better than zinc coatings. For example, its corrosion products are less voluminous, so that there is less danger of cracking of bolted joints. Cadmium is also endangered less by white rust, especially when this unsightly and non-protective corrosion product is produced by action of organic vapour (formaldehyde, formic acid or acetic acid).

5.5.3 Aluminium coatings on iron

In contrast to zinc, the inherent passivity of aluminium as a coating metal is maintained in most atmospheres, and is not upset by high atmospheric sulphur dioxide concentrations. In exceptional cases, especially when there is an extremely high chloride content in the air, the passive nature of the coating metal may be disrupted locally, but this never leads to the electrochemical activation which would be necessary for long-distance protection. Aluminium coatings are noble, and so retard rusting at defects (e.g. pores) in the coating only to a limited extent, and this mainly by a caulking action by 'mixed' corrosion products of the substrate and coating metals. Like zinc, aluminium can repress rusting by binding the stimulating anions into sparingly soluble compounds [22]. The pore-caulking effect is particularly noticeable on metallized coatings; if thin thermally-sprayed aluminium layers are exposed to the atmosphere, there is often an initial showing of small rust-points, which vanish later to leave the coating with a dull uniform greyish-white appearance. Theory and experience both suggest that no protection of cut edges may be expected on hot aluminized sheets. Methods of producing the coating are more important for aluminium than for zinc. Thermal spraying is the most widely used process, but it gives good results only if the layer is practically pore-free or has only self-sealing pores.

Hot–aluminized steel is seldom exposed to aggressive atmospheres without being given a further protective coating. The Sendzimir process is virtually the only one available for this, and produces coatings which are usually not pore-free. Since aluminized steel is produced in rolls, it is again necessary to consider the problems of protecting cut edges. Thin flaw-free aluminium coatings can be prepared by vacuum deposition [26]. Use of this for protection of steel components is only rarely possible, however, since the layers, which are only a few μm thick, are very easily damaged. Only in cases where the danger of mechanical damage is very small (e.g. small steel springs in precision instruments) can the benefits of the good protection afforded by these coatings be used in technology.

5.5.4 Tin and lead coatings on iron

These two low melting point metals are only rarely used for protection against atmospheric corrosion, though lead coatings, in particular, have very interesting properties. Both of these metals are more noble than iron, and corrosion products formed on them can only seal pores or other flaws in the coating to a very limited extent. It is therefore absolutely essential that the coating be applied sufficiently thickly to guarantee freedom from pores. The ability of lead coatings to bind atmospheric sulphur dioxide as sparingly soluble lead sulphate is the reason for the very high corrosion resistance of lead coatings in industrial environments.

5.5.5 Nickel and copper–nickel coatings on iron

Protective coatings of nickel, with or without an underlayer of copper, are often used in the electrical and precision tool industries. Nickel remains passive in most atmospheres in which these coatings are used. When particular stimulators, such as sulphur compounds or large amounts of chloride, are present, which endanger the passivity of the nickel, it must be supplemented by a layer of a more resistant metal (e.g. chromium or gold). Nickel (or the copper–nickel system) can inhibit the rusting of iron only if the coating allows no exposure of the iron substrate surface. The passivity of the nickel and the electrochemically noble behaviour of the copper rule out electrochemical protection of the underlying iron.

5.5.6 'Decorative' protective layers on iron

The most important purpose of these layers is not the maintenance of mechanical properties of the material, which might be endangered by atmospheric influences, nor long-term maintenance of a desired physical state of the surface (hardness, abrasion resistance, electrical conductivity), but rather longevity of their decorative action. To produce such coatings on steel, the metal which is to be in direct contact with the atmosphere must be either immune to corrosive attack (noble metals such as platinum or gold) or have a highly stable passivity, such as chromium.

Such coatings are widely used in the automobile industry and in manufacture of other machinery. They are produced mainly by electrolysis, and are usually based on the nickel-chromium layer combination, sometimes with a layer of copper under the nickel. Methods of production and economic factors will be ignored in this discussion of corrosion in iron-decorative layer systems, though such factors are extremely important in choosing the coating. Rather, it will be investigated how the reasons for their corrosion resistance affect their long-term decorative function.

The requirement that the coating remain 'metallically bright' under aggressive conditions demands that there is complete freedom from pores, since all metals used in these coatings are more noble than iron, either intrinsically (copper) or because of their passivity (nickel or chromium). The method of preparation must therefore be such that a sufficiently thick layer is applied to ensure this. Formation of rust points on the surface after long or short action by atmospheric effects is not related to exposure of previously covered pores; the rusting follows local corrosion of the copper and/or nickel by an electrochemical process whose mechanism resembles the well-known pitting corrosion of passive metals [23, 25].

The bright passive chromium layer which produces the decorative effect of the coating can be applied only very thinly (less than 1 μm) by virtue of the method used. It is very porous, and greater or lesser areas of nickel remain exposed. Since the chromium remains passive and nickel in the pores is easily activated, intensively active local cells between the chromium and nickel are set up, and the action of these leads to relatively rapid perforation of the nickel or of the underlying copper layer. The result is exposure of the substrate metal i.e. the steel, which corrodes because of the closed-circuit cell with the more noble coating metals, and this leads to outbreaks of rust at localized points (Figure 45).

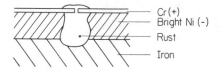

Figure 45. Schematic diagram of pitting corrosion of a nickel–chromium coating on iron.

Activation of exposed nickel is possible. Chlorides are especially effective in this; chlorides are spread on roads in large quantities during winter to act as de-icing agents, and are always present in coastal regions. If there are initially only low chloride concentrations present, accumulation of Cl^- ions by the initiated corrosion process at the nickel anode may be assumed to produce simultaneous acidification. (If it were not an acid medium, nickel would be protected by basic chlorides, so that the pitting would develop more slowly). The frequent occurrence of typical pitting corrosion phenomena on decorative (copper–)nickel–chromium coatings, which is due mainly to the porosity of the chromium layer, has led to the development of more resistant coating systems.

The corrosion rate is determined, other things being equal, by the chromium (cathode) to nickel (anode) surface area ratio. Only pitting corrosion, which proceeds perpendicularly to the surface, is therefore injurious to the decorative properties of the coating. This type of corrosion may be reduced by altering the surface area ratio of chromium to nickel so that the exposed nickel area is the greatest possible. Under cathodic control of the corrosion process, the decrease in anodic current density will then give a decreased rate of pit penetration.

Other methods involve the diversion of corrosion processes which are occurring perpendicular to the surface by applying an anodic 'sacrificial layer', or use of completely pore-free chromium layers. Considerable technical use is made of the first two of these possibilities; the third alternative i.e. pore-free layer production, has hardly been achieved to date because of problems in the process technology. If such layers, which cannot be too thick, are used, their internal stresses, especially those produced by mechanical working, often lead to uncontrollable crack formation.

Special process techniques enable the chromium deposition to be such that, while the appearance (or brightness) is not impaired, 'micro-cracked' chromium layers are obtained whose connected network of cracks leaves a sufficiently large area of nickel surface exposed that the penetration rate by pitting is reduced so far that a longer useful lifetime of the coating is obtained.

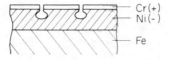

Figure 46. Schematic diagram
of pitting of nickel under a
low-porosity chromium layer.

Figure 47. Schematic diagram of
pitting of nickel under a highly

layer.

A similar improvement of corrosion resistance is caused by microporous
chromium layers. A co-deposition with very fine, non-conducting, normally
inorganic particles suspended in the last nickelizing bath produces the pre-
cursor for deposition of a chromium layer with a denser pore structure.

One or more intermediate layers of non-noble nickel, produced mainly
by high sulphur content (and lying directly below the chromium) help to divert
the direction of corrosion. Instead of penetrative pitting, the corrosion attack
spreads under the chromium layer, and only when the first of the nickel layers
acting as a 'sacrificial layer' is consumed does the essentially local pitting
commence. A further non-noble layer can retard the corrosion by again changing
the direction of the attack. These and similar systems have found very wide

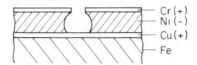

Figure 48. Schematic diagram of a
system to increase the lifetime of a
decorative Cr–Ni–Cu coating on
iron.

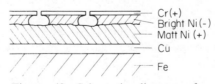
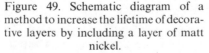

Figure 49. Schematic diagram of a
method to increase the lifetime of decorative layers by including a layer of matt
nickel.

use in recent years; different firms recommend different specific combinations of layers and special processes.

These two methods of improvement of corrosion resistance of decorative chromium coatings can also be combined into one. The nickel layer adjacent to the chromium may be deposited as a non-noble sulphur-containing bright nickel with trapped non-conducting particles. The microporous chromium layer increases, on the one hand, the exposed corroding nickel surface area, and on the other hand there is a favorable change in corrosion direction.

5.5.7   Other metal coatings on ferrous metals

In addition to the metals discussed so far, though far more rarely, other metals and combinations of metals find practical use here. Nickel–tin alloy layers, which can be produced easily by electrolysis, remain passive and bright in most atmospheres which are not too aggressive. Tin–cadmium and tin–zinc alloy layers give a combination of benefits and disadvantages of the alloy components. Tin makes the coating noble, and so the benefits of non-noble coatings which were discussed earlier are lost, but on the other hand the corrosion products now resemble those of tin i.e. they are adherent and less voluminous, and the coating removal rate is lower because of promotion of passivation [27].

Brass layers are used only if no special protective action is sought and there are aesthetic grounds for their use, in non-aggressive environments. They are usually protected in turn by a coat of transparent lacquer.

Antimony coatings protect ferrous metals if they are pore-free. Dip-antimonized steel is resistant even to hydrochloric acid, because of the high hydrogen evolution potential which is produced.

Thick, pore-free coatings can be produced by expensive hard chromium plating or the cheaper matt chromium plating. Hard chromium layers are, of course, completely resistant to atmospheric action. Their use is, however, normally because of factors other than corrosion resistance. Thick matt chromium layers protect iron very well if there are no flaws present.

Chromium-containing protective coatings produced by diffusion processes have a surface which bears no resemblance to that on pure chromium. The chromium content is far above that needed for resistant passivity, however, so that steel parts treated in this way resemble ferritic high-chromium content 'stainless' steels (i.e. not the non-austenitic chromium-nickel type) in their behaviour. The mechanisms of action of other alloy coatings, such as iron–zinc (especially from controlled dip-galvanizing, sherardizing, or metal spraying) or zinc–aluminium (from thermal spraying), may similarly be described as the result of combined, though rather weakened, effects of the individual alloy components. The influence of the non-noble zinc in iron–zinc alloys persists down to relatively low zinc contents, especially with respect to the capacity to bind stimulating anions in insoluble forms [17]. Zinc–aluminium alloys (mostly in the form of metallized coatings) have rapid caulking of inherent pores by highly resistant zinc corrosion products, but there is virtually no long-distance protection because of the high aluminium content [28].

Other metallic coating species (silicon, molybdenum, titanium, tantalum, etc.) need not be discussed with respect to atmospheric corrosion resistance, since they are rarely considered for this.

Combination of different metal layers to improve the protection of the iron substrate are only seldom used. Under certain conditions, benefits come from e.g. the combination in the iron–zinc–aluminium system. The non-noble zinc applied to the iron acts, though perhaps in a limited fashion, to give long-distance protection, while the aluminium reduces the danger of corrosion product formation because of its passivity. This combination has also proved worthwhile as an e.g. thermally sprayed coating where no sand-blast derusting has been performed, or where the steel has undergone fresh atmospheric corrosion attack after such derusting [18]. The thin zinc coating in such cases can stop further rusting (by binding the aggressive anions), but the essential protection is taken over (after some weeks) by the thicker aluminium layer. Thin zinc–aluminium coatings produced by vacuum deposition on spiral springs have proven similarly resistant [29].

5.5.8 Metal coatings on non-ferrous metals

Metallic protective coatings on substrates other than iron should be discussed briefly. Non-ferrous metals must usually be protected for maintenance of their surface properties (appearance, gloss, electrical properties) and only exceptionally against the danger of impairment of mechanical properties by atmospheric action.

Leaving aside the danger of structural corrosion phenomena, which arise under certain conditions on zinc and, especially, aluminium alloys, metallic coatings are of only decorative use. Plating of tempered rolled aluminium alloys with pure aluminium has been performed to lessen the danger of grain disintegration. Pure aluminium plated sheets are an important material in aircraft construction, because they can be oxidized electrolytically; there is no significant diffusion of alloying elements (mainly copper) into the plated layer.

Forged products in these alloys are usually intended for use where appearance is important. Invariably, (copper–)nickel–chromium coatings are considered for protection here. The ideas on the iron–copper–nickel–chromium system which were discussed earlier apply even more where such coatings are applied to less noble metals than iron, such as zinc or aluminium alloys. Voluminous corrosion products from both of these types of substrate, which form very rapidly after perforation of the coating, can 'spring off' the coating in a short time and cancel its action. The quality of the coating is thus dependent on the thickness, porosity, composition and ability to withstand pitting attack.

Similarly, copper and copper alloys are protected against atmospheric attack by applying a metal coating only if the corrosion products may be visually or functionally injurious. The normal coatings used here are nickel or nickel–

chromium; the demands on quality (i.e. resistance to pitting corrosion) are not so great here. When copper corrosion products are formed, their properties (good adhesion, small volume, low formation rates because of the low electro-chemical driving force) do not seriously impede the protective action of the coating. Zinc or cadmium coatings are only rarely considered for protection here; they are usually employed where there are problems of contact corrosion or, more often, solderability, etc.

Other protection methods are also used in the electronics industry under some conditions. If a noble metal is used as the external layer, this ensures that there is no contact resistance caused by corrosion effects. For these reasons, nickel layers are often gilded or covered with a copper–silver–rhodium (or –platinum) coating. This is intended to protect the coating rather than the substrate, so that desired physical (electrical) properties are maintained even under aggressive corrosive attack.

5.6 Protection by organic coatings

Paints, and in the last decade plastic coatings, are the most widespread method of protection against atmospheric corrosion of metals. Approximately 90% of all metal surfaces are covered with organic coatings. The multiplicity of paint types, of available colours, of application processes, and of possibilities of combining paints with metallic coatings has doubtless led to the importance of this type of protection.

Though paint layers have been used for centuries for protection of metals against atmospheric corrosion, the mechanism of the protection is still not completely explained.

5.6.1 The barrier effect

Organic coatings are used in layer thicknesses of 0·05 to several millimeters, though most are in the range 0·07 to 0·20 mm. Such thin layers of an organic polymer, whether without or (more frequently) with inorganic pigments and fillers, cannot prevent the penetration of two of the three important atmospheric components with respect to corrosion, namely water and oxygen. Isolated paint films i.e. films not sticking to the metal, usually have significant per-meability to these two uncharged species [30]. On the other hand, it can be shown that such films stop most penetration by charged species, such as the anions which are also important in initiation and continuation of corrosion reactions. This can be easily understood; the organic binding material becomes negatively charged when in contact with water because of its mainly negative free-ion groups (hydroxyl, carboxyl) and so can pass only cations. However, the maintenance of electroneutrality and the basic principles of the Donnan equilibrium require that cations may not diffuse alone, and so ions enter the film only extremely slowly [31, 32]. The hampering of ion transport by organic films is not due entirely to the action of the ionic groups attached to the binder;

frequently, the inorganic pigment or the filler is able to react with the ionic species in question (usually the anion) to form stable or insoluble compounds. In this way act, for example, zinc white (zinc oxide) and pigments which contain other basic components which help to form insoluble (basic) chlorides or sulphates e.g. chromate pigment containing zinc oxide as a reactive agent, or red lead, whose basic component (PbO) reacts with sulphate to form slightly soluble lead sulphate. The binding of corrosion-promoting anions and the inhibition of their penetration to the metal surface are thus important factors in the protective action of paint layers.

Though, as was discussed earlier, loose paint films are easily penetrated by water and oxygen, paint films adhering to the surface retard the rate of arrival of these two species at the reactive surface of the metal to a limited extent. Wirth and Machu [30] have shown that there are significant differences between properties of loose films and those adhering to the surface. The adherent films, when in contact with an electrolyte, show an order of magnitude increase in resistance to permeation compared with similar unattached films. It thus seems that the adhesion of the film plays a very important role. The adhesive forces must arise from the adsorptive or chemical (chemisorptive) binding of the two components i.e. the adherence should be controlled by the degree of wetting of the metal surface by the organic components of the paint and by the binding energy. As would be expected, polar and well-wetting paint materials are more adherent than non-polar ones. If water and oxygen penetrate the paint layer easily, the energy barrier of the adsorption force must still be exceeded for corrosion development. This only occurs if this reaction barrier is lower than the free energy of the corrosion reaction; this might be due to low adhesion energies or low degrees of wetting. From this point of view, paint–metal systems should be selected to obtain the highest possible adhesion force. The following properties influence this:

polarizability of the paint material;

its ability to wet the surface (i.e. surface activity);

and the capacity of the metal surface to accept the particular paint material. Surface layers which are inert to reaction cannot produce good adhesion of polar paint layers. For this reason, phosphating, for example, is more than a simple surface maintenance treatment, in that it causes a roughening-up of the surface because of the microcrystalline structure of the phosphate.

The barrier effect changes with time. The aging of the binder because of atmospheric effects leads to loss of its properties. Photochemical reactions, which are often accelerated by photoactive effects of the pigment (e.g. 'chalking' of paints with zinc oxide or titanium dioxide as filler), cause loss of the binder, making the paint layer thinner and more penetrable, until at some thin point the substrate metal is exposed. The barrier effect is similarly altered by crack formation. The capacity to inhibit penetration of aggressive anions may be lost by reduction in the number of free ion groups or by saturation of the basic pigment components, and this also leads to loss of the protective ability of the paint.

The barrier action of thick layers of organic coatings (such as asphalt or bitumen layers) differs from that of thinner layers mainly in the enhanced exclusion of uncharged water and oxygen molecules. If the coatings are free from flaws, water from e.g. atmospheric precipitation penetrates during damp periods to a certain depth, but escapes from the coating during dry weather. As long as there is this restriction of diffusion right through the coating, it is completely protective. The permeability of the layer increases with time, however, due to aging.

5.6.2 Corrosion inhibition by paint layers

In addition to the barrier effect discussed above, the protective ability of the paint layer is significantly affected by its inhibitive action. Obviously, paints which give good protection against corrosion must inhibit the penetration of corrosion-promoting charged or uncharged species, and must simultaneously deactivate the components of such species which do eventually reach the neighbourhood of the metal surface by an inherent inhibiting action.

The modern theory of inhibition of corrosion processes is based on concurrent action of inhibiting (in the sense of passivating) and stimulating solution components [33]. If all chemically-active pigment particles which could bind the anions which penetrate the paint layer into soluble salts are used up, or if the selective ion exclusion is destroyed by ion exchange or aging of the paint, or if pores or cracks form in the coating, stimulating anions can reach the surface. In these cases, a sufficiently large 'source' of the inhibitor must be present to avoid primary adsorption of the stimulator. The two protective effects of the paint layer complement each other.

Surface-active inhibitors in the paint layer help on the one hand to improve the degree of coverage of the substrate surface, and are, at the same time, useful as adsorption inhibitors. An example is the lead soaps, produced by oxidative fission of C_{18} fatty acids to give dicarboxylic acids, which occur in red lead–oil paints. Zinc soaps are also important, though less so. Their protective action may be described as follows:

The adsorbed inhibitor retards the movement of Fe^{2+} ions into the electrolyte, and at the same time causes formation of insoluble, stable ferric compounds [34]. As was shown in Chapter 3, the iron dissolution chain reaction needs involvement of ions such as Cl^- or SO_4^{2-}. The mechanism of this action of the soaps is possibly exclusion of these anions, so that passivation processes are favoured.

If the binder-pigment system is not capable of wetting the surface or is not able to develop high adhesive forces, as happens when e.g. low-polarity binders are used, the passivating inhibitors become particularly effective. Water or corrosion-stimulating species which penetrate the film are rendered harmless by sufficiently high inhibitor activities. Pigments containing chromate ions are particularly active in this respect, but other passivation–promoting anions such as molybdate, tungstate or phosphate also give good results. It is not possible to use relatively soluble salts in corrosion-protective paints; on the one hand, the 'source' of the inhibitor would soon be leached out, and on the other hand the solubility of the pigment must lead to undesirable water uptake due to osmotic forces. Inhibitors are therefore invariably present as slightly soluble salts in the paint, with their solubility controlled predominantly by the chemical composition of the salt (especially by relative hydroxide contents of basic salts). Zinc chromate (basic or normal), barium and strontium salts of chromic, tungstic and molybdic acids, and other passivating anions are examples of such inhibitors in paints.

The pigment cations also play a certain role in corrosion inhibition. The passivation mechanism affects mainly the anodic processes in corrosion. Complete passivation is ensured not only by the required oxidation potential but also by simultaneously holding the activities of passivity-disrupting anions at the metal surface to 'below-critical' levels. Though the paint layer is largely impenetrable to anions because of its negative charge, stimulating compounds can reach the surface of the metal through flaws (cracks, pores, edges) in the layer, and can then disrupt the passivity and cause pitting corrosion. Cations should assist in converting the anions into stable insoluble compounds. Zn^{2+} is a good cation in this respect, since it gives sparingly soluble salts with chloride and sulphate ions (which are the two commonest stimulators). Lead, barium and strontium-containing passivating pigments are similarly very effective, especially in atmospheres containing sulphur dioxide, where they react to yield slightly soluble sulphates. There may also be some contribution from action of these metal ions as 'cathodic inhibitors'; one reason for this could be blocking of the cathodic part-reaction by precipitated insoluble salts.

The work of Evans [35] shows that other pigments which do not give direct passivation by reduction of the oxidative ability of the solution are also effective as inhibitors. Good results are obtained with, in particular, zinc phosphate and the slightly soluble phosphates of barium and strontium [36]. It is assumed that the barium phosphate reacts with sulphate which exists at the surface (e.g. in a sulphate nest on iron) so that slightly soluble iron phosphate and barium sulphate are formed.

$$\text{Ba phosphate} + FeSO_4 \rightarrow \text{Fe phosphate} + BaSO_4 \qquad (5\text{-}1)$$

The phosphate ion itself is an important passivation-promoting inhibitor which is involved in the reaction mechanism [37]. After adsorption of phosphate ions, a thin ferrous phosphate layer forms, and this is in turn converted rapidly to a passive layer composed mainly of Fe_2O_3. (Benzoate (Chapter 5.3) acts similarly).

It appears important to introduce at this point further data on the protective abilities of lead-containing pigments, which remain unsurpassed to date. The classical example here is red lead, Pb_3O_4; others are calcium plumbate, metallic lead (and also lead in partly oxidized forms), and, more rarely, lead cyanamide, $PbCN_2$, and basic lead sulphate. The first two, in particular, are successful inhibitive pigments. Lead chromate and other lead-containing passivating pigments are not regarded as belonging to this group, since the main function of the Pb^{2+} in them is regulation of pigment solubility.

Red lead contains, on the one hand, PbO_2 as an oxidizing agent, and on the other hand strongly basic lead(II) oxide. Both participate in the protection. It is well-known that red lead gives the best results when in combination with an oil-containing binding medium. The basic lead(II) oxide gives water-insoluble lead soaps as primary products of reaction with the fatty acids found in linseed oil. Oxidative fission of these acids (perhaps related to the oxidizing PbO_2 component) yields lead soaps such as lead azelate, $(CH_2)_7(COO)_2Pb$, which are excellent inhibitors [34]. By excluding stimulating anions and inhibiting ferrous ion diffusion, they help to form a stable passive layer composed mainly of ferric compounds. Like the protective action of phosphate or benzoate, the involvement of an oxidant (which is not, on its own, sufficient to produce passivation) is indispensable. Atmospheric oxygen, which is always present because of the high permeability of paint films, serves this purpose in most cases. The PbO_2 component may also be involved.

Since the protective action of red lead is attributable chiefly to its PbO content, it may be questioned why litharge is not a successful pigment for corrosion protection. The lead soaps formed promote an increased water uptake by the paint, which thus becomes liquid at very high soap contents, and this means that the important mechanical properties of the paint layer are lost. Red lead acts less intensively in this respect, and the undesirable flowing effects can be reduced further by partial replacement of the active pigments by chemically inert species such as Fe_2O_3, TiO_2, Cr_2O_3, etc., without loss of protection. The extra protective action of red lead in binding SO_4^{2-} ions which act as stimulators has already been discussed.

Yet another theory of the action of red lead has been postulated. J. d'Ans [38] has shown that on iron surfaces covered by red lead paint layers, metallic lead is laid down by cementation. It may be assumed that this occurs particularly on those surfaces, where the cathodic part-processes are proceeding. It is known that lead inhibits the electrochemical reduction process because of the high overpotential value which it causes, so that if the corrosion is inhibited it must be by suppression of the cathodic partial process. The validity of this theory still remains to be tested. Lead cementation certainly occurs, and is favoured especially by presence of chloride ion. For this reason, the use of plumbous compounds in paints may be dangerous on the very non-noble light metal alloys, since such compounds may, under some circumstances, increase the corrosion of the metal substrate.

The inhibition of corrosion by metals which have higher overpotentials for the cathodic part-reaction is an interesting phenomenon. It is worth considering whether, in this respect, other metals such as antimony or arsenic or their compounds might not also act favourably. Antimony is particularly attractive because it can form stable sparingly soluble salts with chloride or sulphate [39].

Calcium plumbate, Ca_2PbO_4, is a modern, effective, corrosion protection pigment which contains no plumbous compounds. The PbO_4^{4-} anion should, in terms of the theory of Cartledge [37] on the inhibition by comparable anions, inhibit diffusion of ferrous ions away from the site of the anodic reaction, as phosphate does. They should form a thin Fe_2PbO_4 layer, which is converted by oxidation into a Fe(III)-containing passive layer. The plumbate ions may possibly also be reduced to PbO by ferrous ions, to allow the effects discussed earlier, while the calcium ions could cause cathodic inhibition by formation of insoluble calcium salts.

The other lead-containing pigments, such as lead cyanamide and metallic lead (or the partly oxidized form, lead suboxide), act similarly. Lead cyanamide, $PbCN_2$, gives ammonia on wetting, and so makes conditions more basic. Though this can promote the tendency to passivation, it is not always desirable because of the danger of soap formation.

Metallic lead and partly oxidized lead pigments obviously protect by forming lead soaps by reaction with carboxyl groups on the binding medium in the presence of atmospheric oxygen. The filtering-out of aggressive anions occurs similarly to the analogous effect found with divalent lead.

The benefits and disadvantages of lead(II)-containing pigments are well-known: their protective action is rarely surpassed, especially if the pigment is applied in an oil-containing binder, but there are some disadvantages. The heavy lead pigments settle out (and hence the attempts to develop 'non-setting' types). Since the lead acts as an oxygen carrier in the drying (polymerization) of the paint, the usefulness of these pigments in presence of sulphur compounds (especially atmospheres containing hydrogen sulphide or carbon disulphide) is limited. A surface film of lead sulphide may not merely delay drying of the paint; it may conceivably completely prevent drying. A further, very important, disadvantage is the toxicity of lead compounds; because of this, lead-containing paints should in general not be applied by spraying.

5.6.3 Metal-rich paints for corrosion protection

This type of paint has found increasing numbers of new applications over the last decade. This is basically because it supplements the different metal coating processes and has technical and economic advantages, especially if large areas are to be covered.

The zinc-rich paints are the most frequently-used of this group. They contain, in the dry state, more than 90% zinc, and their protective action is comparable with that of other zinc coating systems. Apparently there is no important role played by the (usually synthetic) oil-free binder in which they are applied. The quality of the powdered zinc pigment is very important, since metallic contact must be made between the individual zinc particles.

Aluminium-filled organic paints also afford good protection under certain conditions. Because of their passivity, aluminium paints act mainly as barriers, especially when aluminium flakes, which can themselves form a coherent layer, are used as pigment. Similar results are obtained using flaked mica as pigment.

Zinc and aluminium are also used as mixtures (e.g. 10 to 15% zinc in aluminium). The zinc helps to give the coating a less noble potential similarly to its action in thermally sprayed alloy or mixed layers, so that, especially at the beginning of the exposure, a limited long-distance protection is available. The coating is rapidly sealed by the solid zinc corrosion products, and is less permeable than an ordinary aluminium-enriched paint layer.

Iron-rich organic coatings (usually primers) are useful mainly for priming of rusted iron surfaces. Corrosion of the iron powder, which is present in great excess, can be used to render harmless the stimulating anions in the rust on the iron surface. The original rust and that which is newly formed contain large amounts of iron in lower valencies because of the oxygen deficiency, so that the coating is sealed and its adherence is high.

In contrast to these latter paints, metallic paints have now been developed which are based on non-rusting steel powder in an organic binder. These act basically only as impermeable passive layers.

5.6.4 *Effects of metal pre-treatment on the protective abilities of paint layers*

All discussion up to this point on the protective effects of paints has assumed that the paint will be applied in direct contact with a technically pure metal surface, which has been prepared by pickling, sand-blasting, or some other process. In practice, it must often be decided whether the ever-present corrosion products, or artificially-produced conversion layers (which are thus artificial corrosion products), will affect the protective ability of different paint types. This problem has special economic significance for surface treatment of ferrous metals, such as steel structures.

It frequently happens that steel components or half-finished products spend longer or shorter periods in aggressive environments, so that if no temporary protection or priming is applied, the iron will rust. This problem is also significant in the repair or renewal of paint coatings on rusted steel structures. Galvanized or aluminized steel is often only painted after corrosion products of the coating metal have been allowed to form.

It may be deduced from the discussion in Chapter 3 on the mechanisms of formation of atmospheric corrosion products on the various metals that, more than any other, a mixture of rust and soluble species (salts) is a dangerous substrate for paint layers. This is due to three main reasons:

1. Shielding of the iron surface, which cannot come into direct contact with those paint components which confer protection (giving impaired adhesion and diminished action by the inhibitors present.)

2. The voluminous structure of the rust.

3. Most importantly, enclosure of water-soluble rust components under the paint. As was discussed earlier, all paints are more or less permeable to water and oxygen. Corrosion stimulators can therefore act under these layers.

If, for example, as is the rule in rust formed in industrial atmospheres, ferrous sulphate (perhaps in the form of nests) is trapped under the paint, the penetrating water and oxygen cause continuation of the oxidative hydrolysis, which leads to new rusting and regeneration of the stimulator. Rusting thus progresses under the layer, which may be eventually completely destroyed because of the rust-filled blisters which form under it. Soluble hygroscopic species are also dangerous when caught under the paint layer, even if they do not promote corrosion themselves. Soluble phosphates or chromates (e.g. from incomplete rinsing during the passivating process) under the paint layer can cause formation of water-filled blisters. Osmotic pressure gradients promote water diffusion through a paint layer to an iron surface bearing rust formed in an industrial or marine atmosphere. Only exceptionally does rust formed during atmospheric corrosion not promote corrosion under paint layers; this occurs only in very pure atmospheres [40, 41].

The soluble salts accumulate mainly at the iron–rust interface, where they are bound extremely tightly. Though loose rust may be removed by scraping, brushing, etc., most of the stimulator species remains at the iron surface, and cannot be removed even by washing to dissolve it. This is because there is a membrane controlling diffusion of the chloride or sulphate ions out of the aggregation, and this hinders direct water–salt contact [42].

In this connection, a short discussion of so-called 'rust stabilizers' (rust converters) and of priming paints seems appropriate, because of their aim of 'stabilizing' the rust. The intention of these is to convert the hydroxide rust components into 'stable' magnetite. Analogies with protective magnetite layers formed on iron at high temperatures (e.g. in vapour oxidation of boiler equipment) are, however, not valid. In the latter process, the magnetite formed is strongly bound to the substrate (by epitaxy) because of its formation mechanism; secondarily-formed magnetite, on the other hand, gives a loose intermediate layer under the paint layer which is no more desirable than the less adherent but similarly stable rust hydroxides (α-, γ-FeOOH) [43]. The conversion of the rest of the rust, which contains a greater amount of soluble species, must be dealt with in other ways; this essentially involves making the soluble species harmless by converting them into insoluble species. There are basically two ways of doing this: involvement of the anion in insoluble compounds (with simultaneous binding of Fe^{2+} ions), or the use of the principle discussed earlier of electrochemical 'extraction' of these anions [17]. In each case, the conversion product should contain iron predominantly in the form of trivalent crystalline, and hence stable, products (α-FeOOH) and insoluble compounds (salts) of the stimulating anions. Sulphate is best bound as barium sulphate, while chlorides must be formed into less soluble basic chlorides or oxychlorides.

It is therefore unlikely that long-lasting success will be achieved using a single preparation, based on tannic acid for example [44]. The chemical conversion of rust is a very difficult problem in general. The sulphate or chloride nests lie relatively deep in the iron, while the chemical conversion reaches only to the surface. It is debatable whether a membrane of barium sulphate or iron tannate is a better barrier to penetration of sulphate to the nest than the existing one of iron oxyhydroxide.

Good results have been achieved in this area by thermal spraying of a thin zinc layer onto the steel surface [18]. There are obviously two simultaneous effects here: the kinetic energy of the zinc particles enables them to penetrate into the sulphate nest, while the non-noble zinc in contact with the more noble iron or with electron-conducting corrosion products 'extracts' the anions, and binds them as innocuous, slightly soluble basic salts, which seal the coating. This process allows control of the rapid rusting of freshly sand-blasted iron surfaces which often occurs when treatment must be performed in the open air. The thin thermally sprayed zinc layers make the rust harmless, and are also a good foundation for later application of paint layers.

Soluble salts, especially those which stimulate corrosion, which occur on the surface are thus the greatest danger to the protective ability of paint systems. Cases of paint damage are frequent if there is improper rinsing after paste de-rusting, pickling, etc.

Chemical pre-treatment methods have come into use as an answer to the problem of finding methods for rust stabilization. These methods give solid insoluble products. Phosphatizing is the most frequently used of these methods. The phosphate anions produce a fine crystalline compound layer (with iron, zinc or manganese as the cation), which may be of any desired thickness. This pretreatment is a guarantee of complete absence of soluble salts. The water diffusing through the paint layer may be used by the deposited phosphate to produce further inhibiting action. Phosphate is itself an effective anodic inhibitor and, similarly to zinc ions, can 'heal' existing corrosion nuclei. For this reason, there is greater resistance to under-rusting of phosphatized substrates under cracked or mechanically damaged paint layers. It must be noted, however, that over-thick phosphate layers can impair adherence of paint layers under some conditions; this is easily understood. Mere mechanical fastening of the paint to the phosphate layer obviously gives less adherence than would chemisorptive binding. Results are most improved when phosphate layers are used under less adherent paint layers, because of the enhanced contact surface.

The possibility was mentioned earlier of using hot-rolling oxide scale layers as foundations for paint layers. It can be shown in practice that flaw-free scale layers are a good base for protective paints, but that there is rapid under-rusting of the paint if the layer is cracked or porous, as occurs in most cases. This is just as might be expected. The iron exposed in the cracks is a very limited anodic area with respect to the electron-conducting scale layer on the surface. Following penetration of water and atmospheric oxygen through the paint layer, the local iron–scale cell promotes under-paint corrosion. The iron corrodes rapidly, undermining the scale layer (which thus loses its adhesion), and the result is exfoliation of the scale with the paint on it. In addition, the cathodic reaction produces alkalinity of the electrolyte on the scale surface, so that the paint layer is destroyed (e.g. by saponification). Electro-osmotic transport of water also plays a part here. Scale layers can therefore not be regarded as good foundations for paint layers. Cracks and pores, which are formed in the scale layer even with careful handling, lead in the short term to lifting-off of the paint layer.

5.6.5 Protection by plastic and rubber coatings

Plastics are applied to metals as protective coatings by various processes. Some possible methods are tumble sintering, electrostatic powder spraying followed by sintering, dipping into a solution or suspension, stoving, etc. The application of rubber layers is usually by adhesion of sheets and subsequent vulcanizing.

In contrast to paint layers, which are rarely more than 0·2 mm thick, plastics (and especially rubbers) are applied in far greater thicknesses, up to one or more millimeters. At these thicknesses, the main action of the layer is as a barrier.

5.6.6 Oil, grease and wax layers

Again in contrast to paint layers, which are used mainly for long-term protection against corrosion, oils, greases and waxes give mainly temporary protection. (Unpigmented clear lacquers, such as stripping lacquers, also belong to this class of temporary protection agents). The mode of action of these materials is covering of the metal surface to lessen the danger of initiation of corrosion by water condensation. The permeability of relatively thin oil films and of rather thicker grease and wax films for liquid and vapourized water is too high, however, to allow anticipation of good protection in humid atmospheres. If the adsorption forces between coating and substrate are not sufficiently high, the penetrating water will displace the adsorbed film, and since there will almost invariably be sufficient atmospheric oxygen present, the corrosion process can develop. Impregnation of the film-forming material with inhibitors is therefore used to try to produce the highest possible adsorption at the metal–protective film interface.

The protective action of preservative materials based on oils, greases and waxes is thus due mainly to adsorption, and is destroyed by concurrent adsorption of water. Effective protective media must contain a definite amount of adsorption-active additives, and may dissolve only very limited amounts of water, so that the water activity does not rise high enough to allow displacement of the adsorbed inhibitors [45]. It is self-evident that oils, greases and waxes which are to be used for temporary protection must not contain corrosion-stimulating impurities (such as water-soluble acids).

Attempts to improve the protection value of such preservative media by adding passivating inhibitors have met with only partial success. For this purpose, soluble derivatives of passivating acids (e.g. of chromates) can be added to the organic film, so that the water which diffuses in is enriched with the inhibiting anions and the passivity is maintained. Oil- and wax-soluble derivatives of chromic acid may be prepared, but they hydrolyse easily and undergo other changes in composition, so that their use as inhibiting additives to preservative media has not been widespread [46]. Such compounds (especially guanidine chromate) look more promising for incorporation into stripping and soluble lacquers, which are used widely for temporary protection.

5.7 Inorganic non-metallic coatings

Two coating types are differentiated in this category: surface layers produced by chemical reaction, and thus 'artificial' corrosion products, and inorganic coatings whose chemical composition imparts to them a good resistance to atmospheric effects, and which can be applied to the metal in various ways.

Both groups are used widely as protectors against atmospheric corrosion.

5.7.1 Chemically (or electrochemically) formed layers

Inorganic protective layers which are formed by controlled reaction of the metal in different electrolytes are basically insoluble compounds which isolate the metal from the atmosphere. They are often the same as, or similar to, the oxide or hydroxide layers which are responsible for self-passivation of metals, but are artificially made thicker and so are better protectors.

A typical example is the enhanced oxide layer on aluminium. Though aluminium is highly resistant to most atmospheres, it often appears desirable to counteract the occurrence of invisible surface changes. Electrolytic (anodic) oxidation of aluminium components yields a product which is very popular for architectural applications (lining, window framing), and is used extensively in aircraft construction, where both aesthetics and corrosion resistance are important. Layers of uniform appearance which are up to more than $20\,\mu m$ thick can be prepared by this process; these provide a long-lasting stable appearance of the metal after sealing of the pores by partial conversion of the primary aluminium oxide, Al_2O_3, into a similarly very stable hydroxide, $AlOOH$. The porous structure of the primary oxide and the consequent need to seal these pores allows incorporation of a dye during the sealing process. Organic dyes have a good stability here, but other dyes change their tone with time, especially if the surface is exposed to free weathering.

Besides this electrolytic method, protective oxide layers can be produced chemically on aluminium and other metals. Oxidizing solutions, based for example on chromic acid, or merely hot water (in the boehmitizing process) produce thinner, and hence less stable, protective layers which are useful in some applications. In these processes, insoluble compounds of the metal with the corresponding acid anion are formed in addition to the oxide. The majority of technical methods, such as M.B.V. (modified Bauer–Vogel process), Alodine, etc., give mixtures of oxides, hydroxides, chromates, phosphates, etc.

Chromium-containing layers also find wide use. Chromating of (mainly) electrolytic zinc and cadmium coatings produces thin layers which, because of oxidizing reactions, also contain tri- and hexavalent chromium. Chromate layers are fairly insoluble, and their hexavalent chromium content probably forms a source for subsequent reformation of a protective layer when they are destroyed by weathering. These thin layers provide good protection against the frequently undesirable formation of corrosion products, though this protection is short-lived under free-weathering conditions.

The corrosion resistance of other metals can be enhanced by covering them with chromium layers. Examples of these are silver, magnesium, brass, etc.

The problem of surface pre-treatment before painting of non-noble metals can often be solved with the help of low-reactivity oxide layers. It is well-known that reactive metals, such as zinc or magnesium, are protected only with difficulty by oil-containing paints, since the metal soaps which form swell and impair the adherence of the paint. Oxide (or phosphate) layers reduce this danger significantly. Another possibility is the use of reactive primers, which have a mechanism similar to that in oxidation or phosphatizing. These primers contain substances which react with the metal, such as phosphoric acid, as well as chromate-containing pigments and organic film-forming materials.

Iron can also be protected by oxide films. Thin, relatively compact, layers composed mainly of magnetite are formed in concentrated hot alkaline solutions containing oxidizing additives. As long as these films have no pores and the iron substrate is not exposed by mechanical damage, the relatively high stability of magnetite provides protection. Long-term protection, especially in aggressive atmospheres, should not be expected, since the layers are too thin and are easily damaged by chemical or mechanical action.

Phosphate layers on iron, zinc and other metals are undoubtedly very valuable. Their protective ability depends on the barrier action exerted by the insoluble metal phosphates and their mixtures. A phosphate layer is only seldom used on its own as a protective measure, however, since it is not sufficiently mechanically or chemically stable. After sealing with a preservative medium, longer-lasting protection may be expected in mild atmospheres. The most important area of use of phosphate layers is in surface pre-treatment before application of organic coatings, as was discussed earlier.

There are still further alternatives for formation of artificial protective layers using controlled reaction of the metal surface. The low solubility of magnesium fluoride can be used, or it can be combined with chromating. In all these and other cases, the principle of the protective action involves low solubility and uninterrupted coverage of the metal surface.

5.7.2 Inorganic protective coatings

It is convenient to divide these into two sub-groups in discussing them as protective measures against atmospheric corrosion: those produced at high temperature, and those which can be applied at normal environmental temperatures.

Typical members of the former sub-group are the different types of glassy enamels. Thermally sprayed and plasma sprayed coatings of oxides (e.g. Al_2O_3)

are also used, though more rarely, for protection against atmospheric action. The most important area of their use is in combatting hot-gas corrosion.

Glassy enamels are relatively thick coatings, whose most important advantage is an excellent stability against atmospheric effects. Hydrolytic processes which can destroy the glass occur very slowly in all types of atmosphere (including those in which metals are attacked), so that there is practically no loss of quality of properties such as colour or polish over periods of years. The coatings are hard and shiny, and so can be cleaned easily; this explains in part their widespread use in architecture. It is obvious that the enamel must be compact and not expose the substrate. These layers can be prepared easily using suitable methods e.g. the classical two-layer system or the cheaper single-layer method. The brittleness of the enamel layer demands careful handling of enamelled parts, during their erection, for example, to ensure that the protective layer is not damaged; local repair is virtually impossible. The possibility of coating steel and aluminium with silicate enamels is very attractive, since one can thus obtain at a reasonable price a long-lasting stability of the metal, a multiplicity of applications, and varying and durable decorative effects (polish or semi-matt finish, and possible rich colourings).

Inorganic coatings for metal protection against atmospheric corrosion which can be hardened at environmental temperatures remain a rather under-rated area which is ripe for development.

Cement is a very cheap inorganic binder. Certain types of Portland cement can be used, after addition of plasticizers, as cheap but effective protective coatings [47]. If the appearance of the coating is not important and corrosion prevention is the main aim, such coatings seem successful, and they can be applied to surfaces which are damp or have not been specially cleaned (e.g. merely derusted by hand) using a paint brush or a spray-gun with no danger of under-rusting [48]. The protective action possibly comes from the alkaline nature of the cement, which neutralizes the stimulating salts in the remaining rust and in the incoming water and converts them to harmless compounds. It may be deduced from the slowly rising adhesion of the coating and shown by metallography that the plasticized cement coating slowly reacts with the remaining rust [48]. This solid 'paint' layer has proven itself in practice; it finds wide use in Czechoslovakia, for instance, in protection of steel structures in mines.

Zinc silicate paints (with high zinc contents) are another successful type of inorganic coating. The zinc, zinc–lead or zinc–cadmium silicates used as binders weather only slowly. Metallic zinc possesses the advantages already discussed under metal-filled organic paints (Chapter 5.6.3). Since the protective effects come mainly from the sealing action by the corrosion products, it was sought to accelerate the sealing by addition of a more noble metal. Antimony has proven effective here [49]. An interesting variation on this is the inorganic coating developed by Evans, which contains barium phosphate, binder and metallic pigment [35]. This salt should convert sulphate, present in the rust or forming during the corrosion process, into harmless barium sulphate and the corresponding amount of iron into iron phosphate.

Earlier attempts to produce inorganic magnesium or zinc oxide–chloride coatings by direct reaction of excess powdered metal with soluble zinc, magnesium or other chlorides have not yielded the hoped-for protection of steel against atmospheric corrosion, since the binder, the basic chloride, is invariably soluble and weathers relatively rapidly, so that it can promote corrosion [50].

5.8 Combined coatings

Multilayer coatings can provide improved protective action, durability and reliability by allowing combination of different types of coating, and so find many areas of use.

The following multilayer systems should be considered:
1. Metallic + organic coatings.
2. Metallic + inorganic coatings.
3. Inorganic + organic coatings.

5.8.1 Protection of steel by combined metallic and organic coatings

Zinc and aluminium coatings, especially those produced by dipping or thermal spraying, provide good long-lasting protection of steel in most atmospheres even without paint layers over them. However, particularly under the action of atmospheric corrosion stimulators (pollutants, salt spray), the rate of corrosion of the coating metal may become too high to allow anticipation of the desired lifetime of a given coating thickness. Thicker coatings would help here, but these may be difficult to produce for technical reasons (this is especially true of hot galvanizing), or uneconomical. It is satisfactory in such cases to cover the metal coating with a paint layer; the metal coating is then attacked much more slowly, or only after a significant period. Another reason for painting galvanized or aluminized steel is a desire for a particular coloured surface.

In most cases, fairly thin one- or two-coat paint layers are sufficient here; the most important requirement is good adhesion to the coating metal. Difficulties arise in this respect in painting of fresh dip-galvanized surfaces. The use of reactive wash primers or special primers with calcium plumbate pigments help to solve this problem successfully.

There are also advantages in using reactive primers on aluminium coatings, though other paints containing e.g. chromate pigments have proven quite successful.

Thinner paint layers are sufficient here compared with layers applied directly to steel; under-rusting, which is one of the most important reasons for failure of directly-applied paint layers, is now almost impossible. It is relatively easy to choose primers for thermally sprayed aluminium and zinc coatings. The adhesion problem is less important on these because of the surface roughness and the oxides formed during the spraying. Even without inhibiting primers, there is a considerable increase in metal coating life due to the paint layer.

Organic coatings can also be used to protect other metallic coatings against premature failure. Decorative electrolytic copper and brass coatings e.g. on lamps, etc, have a thin transparent coating of lacquer which suppresses atmospheric attack. Aluminium layers produced by vacuum deposition, which are usually only a few tenths of a micron thick, can have their susceptibility to damage reduced and their polish maintained by protection with transparent lacquer layers.

5.8.2 Protection of steel by combined metallic and inorganic coatings

These combinations have so far been seldom used in practice. Good results have been obtained using thin ceramic enamel layers on thermally sprayed aluminium coatings on steel substrates. The enamel seals all the pores and 'copies' the surface profile of the coating; allied with the wide choice of colour and of polish type, this has many possible uses in architecture. The very long service life and easy maintenance balance out the increased cost.

5.8.3 Protection of steel by combined inorganic and organic coatings

Inorganic silicate paints, especially zinc-rich ones, can be coated with organic paints to improve their appearance and to prolong their protective action. Because of the alkalinity of the basic inorganic coating, the organic paints must not be saponifiable. Light rinsing with dilute phosphoric acid before painting of the inorganic layer helps to improve the properties of the system.

The combination of phosphate or chromate layers as substrate for paints or plastic layers, which is used more frequently, belongs more correctly to the category of surface pre-treatment methods, where it was discussed in more detail (Chapter 5.6.4).

6
Principles of Methods for Investigation of Atmospheric Corrosion

6.1   The significance of test methods in atmospheric corrosion protection

As was mentioned in the introduction, the atmosphere forms the most wide-spread corrosive environment. The multiplicity of its corrosion-promoting properties, which depend on the climate and effects of civilization, and its many-faceted aggressiveness to metallic products make design and organization of authoritative test methods very difficult.

Good tests i.e. ones which are fairly inexpensive and rapidly provide comprehensive information, are extremely important from an economic point of view. Since practically all metal surfaces must be protected against corrosive effects of different atmospheres, there are widely varying demands on protective layers with respect to technical and aesthetic factors. It is desirable that there should always be test methods available which will allow optimum choice of the protective coating material and its method of application.

It must also be remembered that the atmosphere, as a corrosive environment, has some properties which change with time, and these may not be simple daily or annual variations. Industrialization introduces new atmospheric influences. Individual microclimates are set up in small enclosed spaces (see Chapter 7), so that changes in manufacturing processes, for example, have an effect. The test method must allow reproduction of the widest possible variations in many variables if precise conclusions are to be drawn. It should also be noted that corrosion tests generally do not have high reproducibility when compared with e.g. analytical chemistry techniques. Selection of test methods in the field of atmospheric corrosion is thus no simple problem. If reliable, meaningful results are to be obtained, the tests must be planned and performed very carefully, and they should above all be evaluated by the most suitable methods.

6.2   Principles of investigation methods

The multiplicity of atmospheric effects discussed earlier, their variability over a period of time, and not least the generally fairly low corrosion rates (compared with those in aggressive environments like acids, etc.) are special problems in evolution of good corrosion test methods.

To obtain accurate, meaningful test results, methods should be used which utilize the knowledge of atmospheric corrosion mechanisms and their kinetic principles, where they are known.

Well-organized natural testing is very expensive, however, and gives results only after relatively long periods, so that it is rarely used in e.g. technical development tests. The chief aim of these tests is the discovery of regularities of natural corrosion processes which will allow development of improved laboratory methods.

The focus of routine testing must be the laboratory. Three types of methods can be used in laboratory investigations, which should be devised to provide data on the behaviour of the metal as well as the long-term and temporary corrosion protection during the test. These are:

1. Model testing. The aim of these tests is the closest possible simulation of natural corrosion effects. Only decisive factors are taken into consideration, and those without effects are stabilized. Accurate evaluation of dominant combinations of factors is not simple, and will be discussed more fully later on. Model testing in the laboratory does not use acceleration of the corrosion, but rather enables exact tracing of the corrosion course and a good evaluation of the changes which occur.

2. So-called short-term testing. This type of accelerated testing should be performed similarly to model testing. Again, the dominant combination of factors which produce the corrosion in the case under test must be singled out. The acceleration is produced by enhanced action of the dominant combination of factors, but this enhancement cannot be unlimited; the mechanism of the process must not be distorted away from that occurring in nature, since this will produce erroneous results. The problems involved in short-term testing are discussed in more detail later.

3. Indirect testing. This type of test holds the most promise. If it is possible to find specific relationships between certain properties of the metal (or metal-coating system) or the atmosphere and the course of the corrosion, very good data on the probable corrosion behaviour can be produced. Some examples of such tests will be described in Chapter 6.4.3.

6.3 Testing in natural atmospheres

As was mentioned in the last section, properly organized and performed testing in natural atmospheres is the best basis for derivation of model tests, accelerated tests, and indirect tests to be carried out in the laboratory. Natural tests are frequently the only reliable basis for comparison of different protection methods, etc.

The relatively slow rate of natural corrosion is a barrier to high rates of technical development. New metals and alloys are introduced, protective coatings are produced from new materials or by new or improved processes, and the engineer cannot wait for several years for comparative tests to be performed under natural conditions. He needs the results much more rapidly, to avoid being overtaken by even newer innovations. The original aim of natural testing has therefore diminished considerably in importance. In addition, the atmosphere has so many variables involved in affecting corrosion properties that long-term comparative testing in natural atmospheres is very expensive. Many organizations must be coordinated in establishing and running the test apparatus in various typical atmospheres, and this often creates problems. Cold alpine climates pose different problems to those in industrial atmospheres, and tropical rain forest climates cause different problems to atmospheres in coastal regions; yet these classifications are not sufficient. Different industrial atmospheres have different types and amounts of emissions, different coastal regions may have different atmospheric salt contents, etc., and there may be general climatic features superimposed on these. Significant predictions of long-term corrosion behaviour in e.g. an individual industrial atmosphere can therefore hardly be made if the basic investigations have been performed only in an atmosphere of a very generally defined type.

What must be taken into consideration, then, if the most important aim of testing in natural atmospheres, which must be the discovery of regularities in the atmospheric corrosion, is to be achieved? The most important requirements may be summarized as:

Suitable choice of the test region.

The most complex possible determination of the important atmospheric parameters, using correctly chosen methods and careful techniques.

Choice of test conditions appropriate to the test; i.e. a suitable configuration of the test apparatus, and a suitable shape of the specimen to be tested.

Sensible selection of specimens and their number, and an accurate description of them and the conditions.

Choice of suitable methods for evaluation of the results of the investigation; these should be as quantitative as possible.

Wherever possible, measured results should be contrasted, so that an accurate picture of the course of the corrosion and of the effects of the atmospheric factors which have been monitored can be obtained.

6.3.1   Choice of test stations

The choice of sites for test stations must be done with a view to the general aim of the test. For example, different concepts are needed in investigations on protection of construction steels to those involved in a systematic determination of the corrosion resistances of protective coatings (paints, electrolytic metal coatings) for use on motor vehicles. The wide range of use of systematic tests under atmospheric conditions demands an especially complex assessment of the greatest possible number of criteria, using equipment at suitable test stations.

To obtain sensible test results, the individual test site should have typical climatic conditions with respect to the area where the system or material under test is to be used. For this, a basic knowledge of the atmospheric factors affecting the corrosion, on the one hand, and of typical reactions of different types of product, on the other, are absolutely essential. Simplified climatic concepts, such as damp tropical or temperature or arctic climates, or marine or industrial climate, can scarcely help if geographical peculiarities are involved. A major role is often played by meso- or microclimatic effects. For example, the author has seen a sea-side test station with stands only 10 to 30 m from the sea, in the tropics, at which the detectable accelerating action by chloride was minimal. The reason for this astounding result is the geography of the area; adjacent to the 500 to 1000 m of beach and flat land, there is a mountain range over 1000 m high, and the cooler air from the mountains flows predominantly towards the sea, and only rarely vice versa.

Other local features can also have a marked effect on climate. In mountain regions, there are known to be great differences in temperature and humidity plots recorded at peaks and valleys, though the altitudes are, at first sight, insignificantly different. Large areas of water also have an effect.

It is very difficult to select typical environmental conditions in urban and industrial atmospheres. Like that of coastal areas, the atmospheric aggressiveness can change significantly in a distance of a few meters, due to action of air pollutants. Thus, not only geographical and climatic effects (topography, prevalent wind directions), but also periodic or non-uniformly varying types and amounts of air pollution must be taken into account.

Exhaustive evaluation of climatic measurements is also very important in the choice of positioning of a test site. If the site is not directly adjacent to a site where meteorological measurements are made, it is advisable to make special comparative meteorological measurements over a long period, to allow some testing of the acceptability of the data from the meteorological measurement site.

Basically, there should be a suitable combination of the dominant climatic factors for the chosen test at the site, and there should be neither very low nor very high intensities of individual factors, or their combinations. For certain programs, test results including occasional extreme values of climatic factors are valuable. With their help, relationships between corrosion rates and degrees of action of the different or combined climatic parameters can be derived, and in this way the possibilities and limits of accelerated laboratory tests can be assessed.

The establishment of a network of test sites will now be described, as an example of the ideas described above on the choice of test sites to obtain a comprehensive assessment of all types of unprotected metals and different protective coatings for general use.

Rapid and reliable recommendations on optimum protection against atmospheric corrosion in temperate climates can be given to the various interested industries, based on results from a series of chief and sub-testing stations. The chief testing site should have a typical temperature humid inland climate. The choice of this site should result from intensive analysis of published and specially-made meteorological measurements; these measurements should include precipitation, humidity and temperature records and chemical analyses of the atmosphere and the precipitation. A site with 'typical' properties is chosen on this basis, at a locality where there is a low concentration of corrosion-stimulating impurities in the air. The properties and results at this chief test site are used as a reference point for comparison of the climatic factors and corrosion activity at the other sites.

Another main testing site is established as a reference point in a typical industrial urban atmosphere. A 'typical' type and degree of chemical pollution of the atmosphere, in comparison with the chief site, is sought. Extremely high pollutant concentrations and non-typical compositions are undesirable (e.g. the air round a fertilizer factory), and so a site near the centre of a town is usually chosen. It is usual at such a site to find sufficiently high concentrations of SO_2 and soluble sulphate-containing dusts present in the atmosphere to allow satisfactory comparisons.

For estimation of corrosion in marine atmospheres, a test site is established on the sea-shore, with provision for test racks to be sited at different distances from the water-line. It should not need to be mentioned that the geographical orientation of the sea-shore must be considered, and that measurements of the atmospheric salt content must be performed.

To answer the frequently-posed problems of corrosion during export to tropical climates, another test site should be established in a damp, warm climate, preferably in a tropical country, with branch sites on the coast and in an industrial town.

This network of basic sites is then supplemented, as occasion demands, by other temporary sites (though these 'temporary' sites may be used for continuous experiments of over periods of several years). In this way, the effects of manifold and extreme conditions in corrosion-promoting atmospheres may be examined and evaluated.

6.3.2 The technical equipping of test sites

For a test site to function satisfactorily, the apparatus must again correspond to the aim of the test. Usually, it involves a simple test stand to which the individual specimens are attached (Figure 50). There are national standards which prescribe the configurations, wind direction orientation, method of specimen attachment, etc., which are to be used during the test. Several points should be borne in mind during construction of the test stand:

1. The effects of microclimates near the ground should be excluded (unless the experiment is determining this effect).

2. Different orientations on the test stand with respect to the ground and the prevailing wind direction should be available (though this can be achieved using a suitably shaped specimen).

3. The stand should permit exposure of specimens with different dimensions or shapes. If the specimen is to be a model or a finished product, special apparatus must usually be used for its exposure.

Figure 50. Test stand for exposure of
samples.

4. The test stand itself should be sufficiently resistant to atmospheric corrosion and other climatic action.

For much data, such stands are sufficient equipment on their own for the test area. More useful results are obtained, however, if in each case suitable meteorological and chemical measurements can also be performed, so that there can be comparison between the test results and the climatic effects during the exposure periods. This also allows more meaningful conclusions to be reached with respect to the life of the material or coating under test, and better comparison between individual test specimens. Methods of measurement and calculation of meteorological (and chemical) parameters are described more fully in Chapter 6.3.3.

Direct outdoor weathering is not the only condition examined at test sites. The corrosion expert is often asked how materials or coatings will behave in the absence of direct weathering. This problem arises not only in long-term applications, but also, and more frequently, in the way it affects the reliability of methods of temporary protection of metals under different macro-, meso-, and micro-climatic conditions. Equipment for testing under simple roofs is therefore often required. Slatted huts appear the best for this purpose, since they exclude direct action of liquid and solid precipitates and of sunshine, while retaining all the other properties of the test site which affect the corrosion (Figure 51).

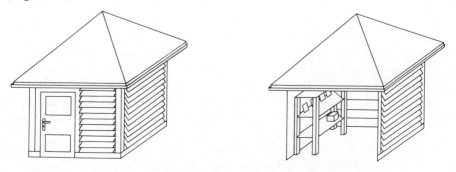

Figure 51a, b. Louvred huts for use at test sites.

A simple but properly equipped laboratory in which the most important measurements at the test site can be performed (with a storeroom for spare stands, specimens, materials, etc.) is indispensable, particularly at larger test sites.

Experiments on actual products are an important facet in practical testing. From such tests, it is possible to implicate manufacturing effects which remain undiscovered in experiments with 'artificial' specimens or models. Cases often occur where the protective surface exchanges heat with the atmosphere during operation of the product. This produces a quite specific microclimate, which can be simulated in natural testing only with difficulty. If, for example, a liquid at constant temperature flows through a pipe-line, this determines the temperature of the pipe-line surface (even if there is an insulating layer). A water-pipe, which is colder than its surroundings for most of the year, is a well-known example. There are similarly examples of heat-emitting surfaces which are never subjected to prolonged wetting.

Experiments under natural conditions using manufactured products as specimens also allow inclusion of other important effects in the test program. For example, in a program examining the optimum protection system for maintenance of steel structures or other rusting objects in a strongly aggressive atmosphere, the influences of different pre-treatment methods on the coating life can be assessed. It is thus possible to examine all important factors (including economic ones), so that the tests can have the strongest possible conclusions as long as they have been properly planned and carefully performed. A further advantage in test reliability is obtained if large specimen surfaces are used.

Results obtained at one site should be used for decisions regarding application at other sites only with extreme caution. This is certainly possible under some circumstances, but only if all the parameters necessary for generalization have been measured; these include meteorological measurements, analyses of air pollutants and precipitates, analyses of corrosion products, surface temperature measurements, etc.

6.3.3 Meteorological-chemical characterisation of test sites

The measurements of chemical and meteorological parameters for this purpose must be chosen on the basis of obtaining information about the effects on the corrosion of important combinations of these parameters (see Chapter 3).

Data describing global climates, measured and published according to standard methods, is demanded mainly by the biological sciences. It is very difficult to use such data directly in atmospheric corrosion problems, since such data do not take into consideration the physical and chemical relationships between climatic effects and atmospheric corrosion phenomena. Long-term average values of temperature, humidity or precipitation tell very little about the corrosiveness of the particular atmosphere. For example, it is not equivalent for continuation of atmospheric corrosion if a daily or monthly average relative humidity value of 60% is produced by humidity variations

128

between 50 and 70% throughout the period, or by intermittent short periods of 90 to 100% relative humidity (and simultaneous precipitation, perhaps) with longer dry periods (say 40 to 50% relative humidity). In the first case, corrosion is negligible or non-existent, but it may be quite significant in the second case. Similarly, data on maximum, mean and minimum temperatures are of little use in corrosion studies, and neither are infrequently-performed analyses of air pollutants and solid or liquid precipitates.

Atmospheric corrosion of metals progresses only in periods when there is a surface electrolyte present, and the rate of corrosion during such periods is related to the corrosion activity of the surface electrolyte and the (chemical) nature of the metal. The meteorological measurements should therefore indicate the duration of presence of surface water. Liquid water may be present on the surface from three sources: precipitation (rain, falling mist, melting snow), dew (when the dew-point is exceeded), or conversion of water vapour into liquid water during a sorption process (when the so-called critical relative humidity is exceeded). All of these processes can be defined by suitably-performed meteorological measurements. Records of annual, and more particularly daily, temperature and relative humidity measurements are an important requirement for this. If these parameters are measured sufficiently often, or continuously (with a thermohygrograph), the time of wetting by a surface electrolyte layer, and hence the duration of active corrosion, can be determined easily [1]. This data allows construction of diagrams which represent the individual periods with hour-of-day as ordinate and month (or a shorter period) as abscissa (see Figure 6) [2]. The lines here represent particular humidities, and using these the duration of any desired humidity value can be determined. Isotherms can be derived and used in the same way.

Direct evaluation of a continuous record of temperature and relative humidity can give similar results. The whole record can be examined and the 'above-critical' sections can be noted; the summation of these gives information on the total time of corrosion (Figure 52).

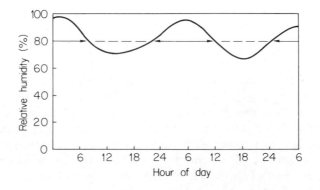

Figure 52. Determination of periods of corrosion from hygrometer measurements.

It is assumed in the foregoing that the critical humidity is known; this must be determined in the laboratory in many cases, by experiment [3]. As was explained earlier, it is dependent on many effects. In particular, hygroscopic substances, from the atmosphere (salts) or formed during corrosion by the conversion of gaseous pollutants, are important in this respect. If the critical humidity has been established, it may be assumed that the time of wetting derived from the humidity record will include the periods of precipitation, which usually occurs only during high-humidity periods.

Under normal conditions under which atmospheric corrosion occurs, a critical relative humidity of over 80% can be expected. Only in exceptional conditions (e.g. extreme atmospheric pollution by chlorine or reactive chlorine compounds) is it lower [4].

The amount of precipitation (i.e. mm precipitation per unit time) is of minimal significance for assessment of corrosiveness of an atmosphere. The measuring instruments normally used in such measurements (pluviograph or ombrograph) are not sufficiently sensitive to measure the vanishingly small quantities of precipitation which are significant in atmospheric corrosion. For example, they do not detect the uniform coverage with dew of surfaces near the ground in weather which is free of precipitation (as such); this phenomenon occurs often, during colder parts of the day.

The duration of surface wetting may be determined directly; special corrosion macrocells have been designed for this (Figure 53). The electrode system e.g. copper–iron or platinum–zinc [5, 6] gives a measurable current in the presence of dew, which can be recorded automatically and evaluated in a similar fashion to the meteorological measurements.

Temperature measurements are important for estimation of the progress of atmospheric corrosion or the aggressiveness of the atmosphere. Annual, monthly, or even daily average values are, like the corresponding humidity values, insufficient. Temperature measurements should be made at the shortest possible intervals (and preferably continuously), so that correlations between temperatures and corrosion can be made from expression of the results as

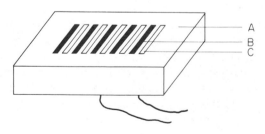

Figure 53. A cell for determining time-of-wetness.
A. Material containing embedded metal electrodes; B. Noble metal electrodes (Pt or Cu); C. Base metal electrodes (Fe or Zn).

isotherms (Figure 5) or by comparison of such isotherm diagrams with times of probable electrolyte presence on the surface. At very low temperatures, for example, the electrolyte may freeze, and the corrosion would then cease. Assessment of temperature results such as these often help in correction of false ideas on the actual temperatures at which atmospheric corrosion occurs. General descriptions of temperate climates include temperatures in the shade of up to more than 30 °C. Active corrosion however, occurs only extremely rarely in Middle and Western Europe at temperatures higher than 14 to 16 °C, and the mean value for active corrosion periods is, in fact, quite close to the meteorological average temperature (which is 6 to 10 °C in Mid-Europe, except in the mountainous regions [1, 7]).

Measurements of surface temperatures are very useful, but are rather difficult to perform for technical reasons; resistance thermometer probes or thermistors can be used here. This sort of measurement is particularly desirable in cases where heat transfer between surface and atmosphere is expected because of the mode of operation. They also permit data on true surface temperatures under direct sunlight to be obtained; this is especially useful for evaluation of tests of organic coatings, whose aging is known to be accelerated by effects of higher temperatures.

Measurements of air pollutants and other corrosion stimulators provide further parameters for characterization of test sites. Investigations of mechanisms of acceleration of corrosion processes by gaseous or water-soluble solid atmospheric constituents (Chapter 3) show that the highest or lowest absolute amount of these stimulators in the air does not have a marked effect on the progress of the corrosion. There is, rather, a clear connection between the progress of corrosion and the total amounts of these species coming into contact with a specified geometric surface. The 'concentration' of the stimulators can therefore be determined by simple adsorption methods, and expressed in terms of weight per unit area per unit time.

Sulphur dioxide and other acid-forming air pollutants can be measured using different modifications of the Liesegang method. The following process, for example, has proven acceptable:

Sheets of filter paper (100×150 mm, and 3 mm thick) are saturated with 10% sodium carbonate solution, dried at 70 °C, and exposed on the test stand in such a way that the atmosphere can circulate freely but rain is kept off. A simple roof over the stand is sufficient for this. Sodium carbonate is especially suitable for use here, since the relative humidity over a saturated solution of it is close to the critical relative humidity which is valid for most important cases of atmospheric corrosion, and so sulphur dioxide will be adsorbed at all times while corrosion is proceeding. The uptake capacity of the filter paper is high. Even in very aggressive atmospheres with extremely high SO_2 contents, the uptake capacity is still not exhausted after a month, and further gas is still adsorbed quantitatively. All the same, the collection period should not be allowed to exceed a month [8].

The analysis of the measurements here is simple and cheap, and uses the classical barium sulphate method. The filter paper is cut into small pieces, leached with hot distilled water several times, and acidified with HCl. Barium chloride solution is added, and the resulting precipitate is simply filtered off, dried and weighed.

Other acid-forming pollutants can be determined by similar methods. There is quantitative adsorption of chlorine and hydrogen chloride vapour, and this can be determined as silver chloride using a nephelometer or by potentiometric titration. It can be shown that SO_2 adsorption values measured in this way agree very well with the amounts of the gas found in rust on steel. The ratio of adsorptive uptake to uptake in the rust is between 0·9 and 1·1, assuming that sulphate is not leached from the rust by rain [9].

Measurement of salt aerosol in the atmosphere is less reliable. The process used most often for this is collection of the salt mist on a pre-dampened surface of filter paper or cotton wool. The surface is dampened by dipping the probe into a bowl of water, which is taken up into the pores in the absorbent by capillary action. The water is analysed at the end of the measuring period. In each case, the prescribed blank analysis for the measurement method should be performed, to exclude possible errors due to inherent sulphate or chloride presence in the adsorbent or chemicals used.

Because of their simplicity of performance and their cheapness, and especially because of their clear relationship to the atmospheric corrosion, adsorption methods for determination of corrosion stimulators in the atmosphere have become very popular. They allow not only a rigorous description of the particular test site environment, but also conclusions on microclimatic effects, such as orientation with respect to prevailing wind direction, distance from the ground, etc.

Absorption processes for determination of atmospheric corrosion stimulators are less satisfactory. They are either only periodically performed (by passage of a defined volume of air through an absorption solution, and subsequent analysis), or performed continuously if automatic recording apparatus (e.g. a polarographic analyser) is available. This latter apparatus is expensive, and requires expertise for its use. Where such measuement results are available, however, they can be converted into adsorption values [7, 8] (see Figure 39).

Analysis of liquid precipitates is not very worthwhile. It is well-known that the composition of rain changes as a particular fall continues (with more sulphate, SO_2, chloride, etc., dissolved at first than later). The amount of stimulator deposited on the surface by precipitation is small in comparison with that from adsorption, the more so in that rain-soluble salts at the surface, where they occur in far higher concentrations than in the rain, are in fact removed by washing away.

Periodic analyses of solid precipitates (dust) are not indispensable, but are nonetheless useful in allowing assessment of possible effects from soluble components of the solid precipitate (mostly salts). The action of dust during indoor exposure can be important, though there is some question whether there are corrosion products formed due to this or whether the temporary corrosion protection methods can prevent it.

A description of the intensity and duration of solar irradiation should also be quoted for test stands on which organic protective coatings are to be tested; determination of hours of sunshine using a heliograph is frequently sufficient here. Energy measurements are more difficult and expensive, and so data from a local meteorological station are usually used for this.

6.3.4 The test specimen

The choice of configuration of the test specimens, their number, and their manufacture are the most important questions to be resolved in any program, once the test site network is established. All of the factors likely to affect the corrosion should be taken into account in their design, and the design of the program. The properties of the metal (particularly its surface state), the overall surface treatment processes, and the construction of manufactured products all play important roles. It should always be attempted to produce specimens which have points of reference to practical samples, and thus to produce and use the metal in a 'model state'. It is often suitable, and of course simple, while investigating the importance of individual factors to hold one or more para-meters constant while examining the remainder. If, for example, the aim of the experiment is the optimizing of an organic protective system for steel surfaces, it seems sensible to examine one series of specimens with identical pre-treatment and different paint layers, and another series of specimens with the same paint on differently prepared substrate metal. A statistical experimental design can often be used to good effect here (though for this, it must be possible to describe quantitatively the different factors, and this cannot always be done).

A description of the specimen including all possibly relevant information is indispensable. Chemical and metallographic analyses of the metal should be available. There should be a detailed description of the surface treatment used, and this should include not only the process techniques but also results of measurements defining the quality of individual operations and of the whole metal-protective coating system. Results of measurements after each treatment stage are thus important; e.g. wettability after degreasing, degree of roughness after sand-blasting, weight or thickness of chemically-produced conversion layers and their uniformity, as well as that of coatings or their components (in multi-layer systems). There should always be accurate information available on the layer thickness, porosity, adhesion and hardness, and on optical and electrical properties of the coating. Not all of the information is required for every test; the selection used will depend on the method of assessment of the test. However, from practical experience, even if it is not always necessary to use all measured values in describing a test specimen, such 'unnecessary' measurements are always less expensive than if the test design has been based on an inadequate collection of initial data. Supplementary measurements are always more difficult to make, even if the 'blank' specimens which are needed are available.

Metal sheet specimens, which are very frequently used, are very simple to produce, manipulate, and evaluate, but they do have some disadvantages. Above all, they are mostly too small a surface to allow observation of effects which are related to the surface dimensions, mainly because of the irregularities introduced to the protective coating during preparation. Other disadvantages in the use of sheet specimens are difficulties in the modelling of metallurgical, constructional and surface-treatment-related processing factors. The structure (or texture) of such a sample frequently differs from the structure of a product made from a similar type of steel (e.g. a rolled section) in quite significant respects, and this can affect the corrosion process under some conditions. Thin metal plates are lighter, and their surface temperature adjusts to that of the surroundings more rapidly, than heavier workpieces, which behave as heat sources and whose surface temperatures only slowly equilibrate with the environmental temperature. This has a significant effect on the time of wetting of the surface, and so on the intensity of the corrosion (see Chapter 3). The configurations of products containing fixing devices (bolts, rivets, welds) is another area of important corrosion effects, which cannot be simulated in a program of tests using flat plates as specimens. There is also a problem in examining benefits and disadvantages of surface treatments such as sand, grit, or chip blasting if thin sheets are used as specimens.

It is clear, then, that the sheet specimens which have been the most popular to date are satisfactory only for limited test aims. They are, of course, ideal for tests such as:

Comparison of the behaviour of different coating systems after standard pre-treatments.

Determination by direct measurement of atmospheric corrosiveness at different sites on different materials (for completion of meteorological-chemical characterization of the site).

Comparison of corrosion rates of analogous materials to assess e.g. the action of different alloying additions.

It is frequently also important to obtain data on the danger of corrosion after possible mechanical damage to the coating, when examining the protective abilities of various coatings. This can be examined easily on sheet specimens, using a prescribed artificial disruption of the experimental coating. This deformation may be e.g. cutting with a knife, or the Olsen cupping test, and this sort of experiment gives information of the most quantitative type possible on this problem which arises often in practice.

Particular test data is often obtained by using a special specimen configuration. If good protective systems for steel structures are sought, it is necessary to simulate characteristic parts of the structure. It is well-known that geographical direction, orientation with respect to the ground, and structural configuration have effects on the life of the coating. Informative results, such as are obtained from simple sheet specimens, are also obtained from shaped sheet specimens which have horizontal, vertical and sloping surfaces, and on which edge effects can be observed (Figure 54).

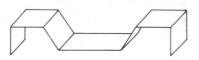

Figure 54. Profiled specimen for
testing of paint layers.

Corrosion protection problems caused by the different fixing techniques cannot be tested without special models. Corrosion in crevices, which are formed particularly during spot-welding, or on riveted or bolted structures, must not be overlooked. A configuration and preparation method corresponding to the actual process are absolutely essential here, especially in cases where limiting values of the effect (e.g. crevice dimensions) are involved in the design of the test program.

The same applies to tests to investigate the corrosion danger from combination of metals in contact with each other, or contact between metals and non-metals. No sensible results can be obtained without taking all factors carefully into account in designing the experiments. The chief factors are the relative sizes of the areas of the noble and base metals which are in contact, and the formation of compounds involving the metal and possible degradation products of organic materials. (This latter effect can be very significant under certain circumstances; e.g. in the enclosed space of a package). The importance of determining all important properties of the materials in contact with one another must be stressed here yet again.

It is difficult to perform experiments in natural atmospheres which will indicate the susceptibility of different metals (or metal-coating systems) to combinations of corrosion effects and mechanical stress. Stress corrosion cracking is not rare under atmospheric conditions (see Chapter 4), but very little is known to date on atmospheric corrosion fatigue. The uncertainties which beset these two fields of corrosion in general also apply to testing of them under atmospheric conditions, and so investigation of them is limited mostly to systems with constant deformation (Figure 55). Experimental methods for testing in natural atmospheres under constant loads are rarely used because of their expense.

It is scarcely possible to enumerate all the points which must be considered during choice of specimens. One more must be stressed, though: the reproducibility of corrosion tests is very slight, because of the extremely high number of parameters which it is difficult to hold constant. The number of specimens for measurement should therefore not be too small. There are some advantages in deriving statistical data based on the spread of results obtained. The number of specimens needed is also related to the method of assessment.

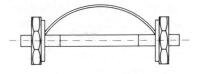

Figure 55. A test configuration for
stress corrosion investigations using
a constant deformation.

The intention of every corrosion test is the determination of the time dependence of the process, and so the states of the samples are examined over a pre-planned period. If non-destructive methods of evaluation are available, which is not always the case, a smaller number of specimens can be employed than if the sample must be destroyed to assess the corrosion. From experience, reliable evaluation by weighing (after removal of corrosion products) needs about five specimens, with total surface area (both sides) of about $0.15\ m^2$. Non-destructive assessment demands samples with total surface areas of 0.3 to $1\ m^2$, depending on the complexity of the effect being tested. When non-destructive tests are used, the sample remains exposed until the end of the test, but destructive methods require a new set of specimens for each period of exposure which is to be evaluated.

6.3.5   Assessment of natural corrosion experiments

In contrast to other engineering fields, corrosion testing techniques are, to date, rarely quantitative, and hence tend to be evaluated objectively. An engineer or architect knows exactly how he can use the quantitatively expressed results of tests of mechanical, electrical or other physical properties of a material, but he is often perplexed as to how to begin to use the verbal descriptions of corrosion behaviour, or non-numeric results of corrosion experiments. Obviously, this last sentence is a deliberate over-exaggeration; tests of physical properties of materials often use unwarranted system simplifications, and frequently ignore the influence of time, and so do not necessarily produce an accurate result.

Test results which state that material A or protective system B behaves better at site X than at Y or Z are certainly valid information, but they do not lead to scientific-technical improvement of corrosion protection methods. To achieve this latter aim, the course of corrosion over a period of time must be determined quantitatively. A properly-planned program will always bear this aim in mind. It seems desirable to divide methods of assessment into qualitative and quantitative types before discussing these methods more closely, and suggesting criteria for their selection.

In quantitative assessment methods, the test results describe the corrosion course in physical units, and embody other data which is important for theory and practice. Measurements of weight loss are very widespread, simple, and useful. The weight loss can be determined easily by weighing the specimen, after removal of corrosion products using standard combined chemical (or electrochemical) and mechanical operations, and the mean metal removal can be calculated. These data are expressed as weight losses in $g\,m^{-2}$, $mg\,dm^{-2}$, or as μm or mm. These values may be converted to corrosion rates, which may be expressed in units of e.g. $g\,m^{-2}\,year^{-1}$ (or day^{-1}), $\mu m\,year^{-1}$, etc. Practical data on corrosion rates should always be examined carefully before re-use; they are usually based on linearized calculations which do not reflect the actual corrosion kinetics (Figure 56). The true corrosion rate changes with time (until it finally becomes constant), as was discussed in Chapter 3, and so it can only be obtained by examination of the corrosion-time curve. Only the values from this curve may be used for extrapolation.

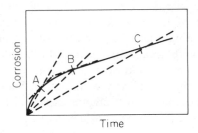

Figure 56. Determination of actual corrosion rates. $---$ incorrect rate; $-\cdot-\cdot-$ true rate.

It is desirable, in some cases, to measure the amount of adherent corrosion product. This may be done easily by weighing before and after product removal; such results provide data which may be of use in e.g. selecting a cleaning process.

The methods of determination of corrosion removal (or weight loss) are very simple, cheap, and give quantitative data. They do have certain disadvantages, however, which can be eliminated by use of supplementary measures. Only rarely is there totally uniform removal of material during atmospheric corrosion. Even the widespread rusting of ferrous metals is not a completely uniform corrosion phenomenon. Similarly, on aluminium alloys, amongst other materials, there are more or less deep 'dimples', corrosion scars, and other irregularities. These can have some technical significance under certain conditions, and distort the validity of mean values of material removal during corrosion. It therefore seems sensible to supplement weight loss values with quantitative descriptions of the uniformity of the corrosion. Mechanical or optical profile measurements (which resemble roughness measurements) are satisfactory here. Corrosion scars (e.g. pits) can be described by their number per unit area, by their depth, and by statistical methods. If all these measurements are performed during the test program, a very comprehensive review of the course of the corrosion can be obtained.

The worth of weight loss measurements is particularly questionable when the corrosion affects structural properties of the material. The danger of grain boundary corrosion is high on many aluminium alloys and certain other materials; in spite of the small weight changes here, the alloy loses its mechanical strength (Chapter 4). The best method of examining the depth of intergranular corrosion and its progress is metallographic analysis. Measurements of mechanical properties which are influenced by corrosion, especially by corrosion which attacks the structure of the metal, also give useful data; natural changes in mechanical properties (due to aging, for example) must also be taken into account in many cases like this. It is difficult to transfer results from one material to another if the corrosion is affecting the material structure, but a statistical study helps to take into consideration the irregularities in non-uniform corrosion phenomena which can act as stress raisers (e.g. pits or notches). Such a study also helps in the understanding of the long-term changes in characteristic properties of the material, which may be individually measured or indirectly derived.

Stress corrosion cracking experiments are usually described only in terms of time to fracture in a given initial configuration. These values have a large scatter, and can be used practically only after statistical treatment.

Other measurable characteristics can also be used for quantitative evaluation methods. For example, if the test allows estimation of changes in surface resistance (for example, of materials used for contacts in electrical apparatus), measurements of this value and its changes with time will give valuable information on probable effects of corrosion products on the capacity of such materials to function in electrical circuit contacts.

Besides data on the course of the corrosion, other data can be collected during tests on metals in natural atmospheres which provides more comprehensive information on the corrosion process. For many purposes, it is sensible to analyse the corrosion products, remembering that they do not have a homogeneous composition. At the boundary between the metal and the products, there are different chemical compositions and structures to those at the product-air interface. Attempts have been made to devise methods which provide information on these differences, which are undoubtedly important theoretical and technical factors [10]. Three different layers can be found in rust layers by e.g. mechanical working of the specimen. The outermost layer is removed by brushing, the middle one cracks off if the sample is bent, and the innermost adheres very strongly, and can be removed only by chemical methods [9].

Chemical fine structure analysis (using X-ray or electron beam diffraction) provides more valuable data for the corrosion expert. Results from thermogravimetric analyses, differential thermal analysis and electron microprobe measurements of the distribution of elements in the corrosion products are also very informative [10]. Optical and electron microscopy (especially using the scanning electron microscope) help in the accumulation of further data [10].

If irregularities of element distributions on the corrosion product surface (such as local sulphate or chloride accumulations in rust [11]) are to be investigated, coloured reaction products from microchemical processes can be obtained which may then be examined microscopically. Electrochemical measurements can also be useful. Changes of polarization behaviour of corrosion product-covered specimens with time allow conclusions to be made regarding the stimulating or protecting action of the corrosion products. The principle of polarization resistance measurement (as a measure of corrosion rate) has not so far been successfully introduced into atmospheric corrosion testing techniques, however.

In particular cases, measurements of optical properties of unprotected surfaces are useful. These are usually used in comparing corrosion-induced changes in decorative coatings, or coatings which require a certain polish or colour for their functioning. They may also be used e.g. to assess the long-term retention of polishes of different degrees on chemically or mechanically polished stainless steel for use as automobile trim. In such a case, measurements of light reflectivity enable quantitative determination of the corrosion rate. Similarly, the progress of corrosion on low-alloy weathering steels (for architectural applications) during weathering experiments can be measured by the rate of change of colour of the rust. This can be done using spectrophotometric reflection measurements, for which there is suitable inexpensive apparatus available [12].

It is far more difficult to select a quantitative method for assessing the corrosion undergone by metals protected by coatings. It is difficult to apply quantitative methods to the metal-coating system, especially in multi-layer systems, to determine the effect of corrosion on the system. Concepts like 'protective ability', corrosion under the coating, and appearance do allow some characterization which can be backed up by numerical values, but almost every protective system (or at least every type of protective system) needs special characterization to describe its corrosion behaviour. The choice of assessment method depends above all on the mechanism of the protective action, and must also be chosen with a view to the technical use of the coating.

Quite different criteria must be applied to primarily protective coatings and primarily decorative coatings. The former are intended to give long-term protection against atmospheric corrosion, and their appearance is secondary to maintenance of mechanical strength of the substrate (e.g. zinc or aluminium coatings on steel surfaces), while decorative coatings such as bright chromium layers are applied to provide a constant appearance. The basic question which the testing sets out to answer is the same in both cases (viz. How long does the coating last?), but the sense of the question differs. In the first case, the test must supply the answer: the coating metal can probably last x years without rust forming. If a longer life of the structure (say y years) is desired, then either the protection of the coating must be increased to cover the period of $y–x$ years (e.g. by increasing the layer thickness, or by combining it with an additional layer), or else it should be expected to have to perform maintenance after, at most, x years. In the second case, the test should give the result: the decorative layer lasts x months without significant change of appearance, without needing maintenance. After $x + y$ months, its brightness has changed (by $n\%$), and after $x + z$ months the appearance of point-forming rust sites (described in terms of number of rust points per unit area) is likely.

Technologists and economists can use this type of result, and this is desirable for all assessment methods.

A knowledge of the mechanism of protection can allow a limited assessment of the progress of corrosion in the protective system by following the rate of weight loss (or removal by corrosion), but it is practicable only for relatively thick zinc, cadmium, aluminium, or lead coatings on steel. The uniform removal of such coatings allows calculation of the time at which there will have been more or less complete removal of the coating, if its original thickness is known. The same result is obtained by applying directly-measured corrosion rates for the coating metal to the corrosion of the coating. The atmospheric corrosion rates of the coating metals are known to be little affected by the method of preparation and the degree of purity of the coating metal.

Since the protective value of a metallic coating can be expressed only rarely as a directly-measured physical quantity, visual observations have to be made quantitative in nature. The way in which the corrosion appears and the technical use of the product are of significance here. There are two methods available: comparison of the changes produced by corrosion with standard samples, or objective measurement of changes. The former possibility gives more rapid results, but does not exclude subjective impressions of the observer. Many national and international standards for assessment of states of corrosion in coated systems are based on use of suitable standard specimens.

It is difficult to make much progress without using objective measurements, however. Bright chromium coatings typically corrode by rusting through at points (Chapter 5). The individual points can be counted for a more quantitative evaluation of the progress of corrosion in terms of points per unit area, and the density of points (which is important for technical reasons) can be examined. This principle is well-accepted in this special area.

Uniform surface changes which are produced by corrosion of the coating (loss of polish, colour changes) can be measured, at least in part, by optical methods, as was mentioned earlier.

Some corrosion phenomena cannot be quantified by any of the possibilities discussed so far; blister or crack formation in organic paint layers are examples. In such cases, a more subjective description of the corrosion must suffice. In most cases, standard specimens are of no help, since these types of corrosion display widely varying forms.

In general, methodology of corrosion protection assessment and evaluation of the resistance of organic coatings is a little-explored field. Too little is known about the mechanism of the protection by the paint layer (Chapter 5), and this makes choice of quantitative evaluation methods difficult. Electrical properties (such as capacitance measurements) have so far allowed no better understanding than does visual assessment [13]. Polarization resistance measurements will perhaps help in the future [14]; these can give reliable data on changes of protective ability of the paint layer.

Evaluation of artificially-deformed–sample testing has still to be discussed. Spread of the corrosion with time (by under-rusting or blister formation) from its starting point at an original cut edge allows a degree of quantification of the progress of the corrosion, and thus a means of comparing the protective value of different systems (including effects of pretreatment).

6.3.6   Selection of assessment intervals

It has already been discussed how corrosion tests must be properly organized to determine the progress of the changes which occur. Only this will allow the collection of reliable data on expected lives of different materials and protective coatings, to give the engineer a basis on which to select optimum protective processes. To determine a meaningful corrosion-time curve, a sensible time plan for assessment of experiments and measurements must be evolved.

It may be deduced in advance from a theoretical analysis of the test system whether the changes which are found during the experiments imply a simple or complex corrosion process. For example, the corrosion-time curve for an unprotected metal (as shown by weight loss measurements) is likely to be a simple curve. On the other hand, the first appearance of a rust point on a chromium-plated product signifies an obvious qualitative change, and only from this point in time can the multiplication of the rust points be described by a satisfactory expression. In the testing of temporary protection methods, the most important information is the time to appearance of the first corrosion phenomenon. The further development of the corrosion will provide extra information, but the first observation of corrosion is the decisive one in this case.

If a simple curve is expected, the test period should be chosen to give the most complete curve possible. An eventual bending of this curve may lead to discovery of theoretically and practically important results, following basic analysis.

If corrosion is continuous, two possibilities arise: either the initial rate is more rapid than the final one, and decreases to reach a constant value after a certain time (which is the case typical of atmospheric corrosion of unprotected metals), or the rate increases with time. This second possibility is found if corrosion is causing a gradual loss of the protective ability of the coating. The case discussed earlier of corrosion of bright chromium coatings is again a good example of this. After a gradual loss of shine, which is technically unimportant, rusting begins. The rate of appearance of new rust points usually increases with time (Figure 57). Protective paint layers cause similar behaviour. There are gradual changes in observed (or measured) properties during the initial period, while the paint protects the metal against corrosion. Only after a certain interval is there apparent corrosion, loss of the protective ability of the

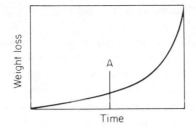

Figure 57. The profile of weight
loss from a predominantly decorative coating.

paint, and increased corrosion rates, until after complete annihilation of the
coating the metal is again freely corroded according to the laws of atmospheric
corrosion of unprotected metals.

Only with the above ideas firmly in mind can a suitable experimental
design be evolved. If a decreasing corrosion rate is anticipated, the evaluation
periods should be in, for example, a geometric progression to obtain the best
results. The following pattern has been proven in practice: 1 week, 2 weeks,
1 month, 2 months, 3 months, 6 months, 1 year and then regularly each year.
This sort of plan is particularly useful if the specimens are unprotected metal.

If sharp changes in the corrosion rate are expected, more frequent assessments must be made as the anticipated time of appearance approaches. The
plan must be designed here with reference to experience with similar systems.

6.3.7 Compilation of data on atmospheric corrosion tests

The large amount of different information which is needed for evaluation of
suitably-organized experiments needs careful preparation. As has been detailed
earlier in this chapter, the following data are involved: meteorological and
chemical parameters of the atmosphere (Chapter 6.3.3), exact characteristics
of the specimen and its preparation (Chapter 6.3.4), and results from intermediate
and final evaluations of the test. The work involved in manual treatment of this
data is quite considerable, especially if there is a large number of specimens (and
numbers of over 10,000 are not exceptional at larger test sites).

Each specimen should have a card corresponding to it contained in a
reference file, onto which can be entered clearly the important data. It is an
advantage to use punched cards, which in turn allows the use of computerized
techniques, and thus rapid sorting and related evaluation of results. Obviously,
each test will require evolution of a special code system for specimen description,
test site information, and evaluation results, and this system must obviously
be related to the type of computer and associated auxiliary equipment which is
available. It is not very difficult to devise such a system. In addition to test site
designation, it can contain all the necessary data on initial properties of the
material under test and of the coating system, the data on the processes used,

and the results of all tests done on the specimen (since a series of assessment methods may sometimes be used) [15].

Climatic data is usually handled separately, and only subsequently matched to the other data. Treatment of numerous meteorological data is a long-winded and hence expensive task. It is advantageous to use measuring apparatus which punches the most important data (temperature, humidity) directly onto paper tape, and so allows computerized assessment of this information [15]. The higher costs involved in this are recovered in every case if the test site is to be established and run for several years. Data on factors which cannot be determined automatically without difficulty, such as air impurity, dust or precipitation analyses, or duration and intensity of solar irradiation, must usually be handled separately. At permanent test sites, it is worthwhile building up a file of punched cards on individual and associated climatic parameters, to ease the later preparation of more exact data on long-term corrosion tests and their correlation with climatic factors.

All of the basic ideas which have been discussed in Chapters 6.3.5 to 6.3.7 also apply to Chapters 6.4.1 to 6.4.3, which discuss methods of laboratory testing.

6.4 Atmospheric corrosion testing in the laboratory

The long periods and high expense associated with testing in natural atmospheres make it sensible to use less expensive laboratory investigations. In laboratory tests, short- or long-term variability of combinations of factors which affect the corrosion, which poses a problem in natural testing, can be mostly avoided. Testing in the laboratory allows the use of rigorously defined conditions, and cyclic changes of the conditions if desired. Design of exact test conditions is not an easy matter, however, if informative results are to be obtained and false conclusions are to be avoided.

As was discussed in Chapter 6.2, laboratory test methods may be divided into three groups: model testing, short-term testing and indirect processes. A closer discussion of each of the three types seems desirable.

6.4.1 Model testing

This type of method should imitate the natural atmospheric corrosion course i.e. 'modelling', without arbitrary acceleration.

The first question to arise here is: What typical corrosion effect should be modelled to provide comparable corrosion paths in 'artificial' and 'natural' tests? It is clear that there is no universally valid answer to this. Our knowledge of the mechanisms and the kinetics of atmospheric corrosion (Chapter 3) suggests that the corrosion of different metals or metal-coating systems are influenced decisively by different combinations of dominant factors. The concept of 'combinations of dominant factors' helps, to a certain extent, to overcome this difficulty. A whole range of atmospheric parameters are practically insignificant with respect to the corrosion process. Neither the air pressure nor weathering nor wind action need be considered, and the amount of rainfall is unimportant in many cases. On the other hand, the following factors, and their combinations, must be correctly reproduced:

Profiles of relative humidity and temperature.

The absolute quantity of stimulator reaching the corroding surface per unit time.

The energy of the solar radiation (though this is important only for corrosion (i.e. aging) of organic protective layers, and needs therefore to be considered more rarely in model testing).

A thorough analysis of climatic and chemical parameters is necessary to discover 'typical combinations of factors' for the corrosion processes in natural atmospheres, so that the atmosphere may be typified. Some divisions which might be suggested are, for example, unpolluted rural atmosphere in the temperate zone; urban atmosphere, polluted mainly with SO_2, in the temperate zone; coastal atmosphere in the temperate zone; damp, warm (tropical) atmosphere with no stimulator action; and so on. Indoor climates may be similarly classified and modelled.

For each individual test case, the duration of periods when the corrosion is active or dormant (e.g. in per cent of weeks, months, or years) and the intensity of the corrosion should be determined.

A suitable approximation to environmental conditions is particularly important for testing models of progressive corrosion. It follows from consideration of the kinetic effects of individual parameters that not every fluctuation need be modelled (which is not practicable, anyway). It is well-known that corrosion of metal surfaces occurs only at above-critical humidities, and is intensive if electrolyte layers form by condensation (e.g. dew) or precipitation. The relative durations of surface electrolyte presence can be determined fairly easily from meteorological data (see Chapters 3.7.2 and 3.8), and it is also possible to divide these periods into parts with electrolyte present because of adsorption and condensation. Adsorption-promoted electrolyte formation is simulated easily by using above-critical humidity (usually 85 to 90%), and that caused by condensation can be simulated by temporary undercooling (e.g. using a Peltier cell), artificial mist generation, or temperature fluctuation. Modelling of typical temperatures during active corrosion periods can also be attained. High relative humidity values and periods of precipitation are generally associated with the lowest temperatures of the daily cycle. For a given climatic zone, active corrosion at higher temperatures is found only exceptionally, and for short periods. Temperature effects on corrosion rates in the atmosphere are not very important (Chapter 3.7.2). Comparison of climatic data with measured values shows that the temperatures during active corrosion periods correspond to the average values for the period (i.e. day, month, year) [7]. For all these reasons, no serious errors are introduced if the model test is conducted at a constant temperature corresponding to the annual mean temperature for that climatic zone.

The action of atmospheric corrosion stimulators must also be simulated with care. It is clear (see Chapters 3.5, 3.7.2, and 3.8) that there is a close relationship between the corrosion rate and the total amount of stimulator which comes into contact with the surface during exposure, and thus the average stimulator concentration in the atmosphere. Again, this amount is a parameter which must be considered in planning the model test program, and achieving a proper modelling is always complicated. The introduction of small amounts of the stimulator (e.g. SO_2 or salt spray) into the test chamber is difficult, though possible, and can be done by periodic dosing or, if the stimulator is a gaseous impurity, by extremely sensitive diffusive methods [16]. As a last resort, a calculated amount of the stimulator (in solution form) can be added at the start of each corrosion period. This is not only time-consuming, but also poses problems regarding the validity of the model test. While it is clear that the anion of the stimulator has the decisive influence, the cation must also be chosen correctly. In modelling of the rusting of iron in a sulphur dioxide-containing atmosphere, ferrous sulphate solution, for example, may be added in known amounts and concentrations. In doing this, however, an important reaction is excluded from the system, namely, the conversion of adsorbed SO_2 into sulphate anions, which probably does not occur uniformly across the surface, so that the distribution of the anion across the metal surface may be distorted compared with the natural case.

Practical simulation of 'salt spray' in model tests is particularly difficult. It is known that the total amount of chloride deposited on a surface rarely exceeds $1.5\,g\,m^{-2}\,day^{-1}$. In spraying salt solution into a test chamber, it is usually impossible to avoid introducing a considerable excess over the desired concentration, which is uniformly distributed throughout the chamber and comes uniformly into contact with the surface being tested. Aerosol generators such as that described by Huber [17] help to overcome this problem to a certain extent, but very dilute chloride solutions must always be used, and values for precipitation onto differently oriented surfaces during the experiment must be determined to allow fixing of the experimental parameters.

Problems of homogeneity of artificial test climates, which are due mainly to the designs of the different types of test chambers and to their technical equipment (temperature and humidity regulators, dosing apparatus, etc.), cannot be examined further here. Regulatory equipment, protected by patents, is available from certain manufacturers, and is the subject of a monograph by Rychtera [18].

Up to this point, only the simulation of active corrosion periods has been discussed, though alternation between these periods and periods of dormancy are characteristic of atmospheric corrosion. This fact should never be overlooked in designing model tests of atmospheric corrosion, since it determines the paths of many of the corrosion processes. For example, there is no significant difference between the corrosion resistances of low-alloy steels with high weathering resistance and normal carbon steels if periods of drying out are not included in the test program.

Long-term testing in artificially-damp atmospheres can lead to quite false conclusions on the behaviour of different paint films. In nature, humid periods during which water penetrates into the paint film are followed by periods of partial (or semi-complete) drying out. In this way, the water uptake of the film scarcely reaches values high enough for irreversible swelling phenomena or blister formation by disruption of the adhesion between the layers in the system to occur. Permanent exposure of paint layers in humid atmospheres or under constant wetting or periodic dew formation conditions thus produces corrosion phenomena which have little in common with those occurring naturally. The results of such tests are accordingly worthless.

For these reasons, it is necessary to include 'dry' periods in the test cycle when devising a modelling test program.

The life of organic coatings is undoubtedly decreased by thermal, photochemical, and other atmospheric effects in addition to those discussed already. Even today, theoretically-based modelling of these effects is impossible, which contrasts with the ideas discussed above which are based on the continuing discoveries of mechanistic and kinetic laws in the atmospheric corrosion of metals. The theory describing atmospheric aging of organic materials is not well enough advanced as yet for it to be used in planning of model tests. The major reason for this is the tremendous multiplicity of chemical properties of organic materials, and hence of the processes by which they may be destroyed in different atmospheres. Data from particular experiments can only be generalized very approximately.

One is therefore forced to conduct tests of protection by organic coatings which are based on observation of effects, rather than on simulation of the destruction mechanisms and their kinetics, as can be used for metal corrosion. It is known from experience that the photochemical action of the energetic UV band of the solar spectrum is much more pronounced if there is simultaneously a high relative humidity present. The sample is therefore irradiated and simultaneously dampened, using a light source with characteristics like those of the solar spectrum (e.g. a high-pressure xenon lamp). Since the surface temperature rises during irradiation due to energy absorption from the IR component of the radiation, this parameter must also be controlled; i.e. the surface temperature should not climb above naturally-occurring values during irradiation.

The total (or the UV component of the) radiation energy per unit surface area per unit time should be another parameter in examining the effect of sunlight during model tests. This is relatively easily measured in natural and laboratory tests [19], and is used in fixing the test program. The value can be regulated in the laboratory by changing the distance between the specimen and the light source, and by measurement of the irradiation periods, while simultaneously stabilizing the surface temperatures.

The effect of frost temperatures should also be taken into account while testing organic coatings. Paint layers are embrittled by frost, and there can be severe damage to paint layers if there is a sudden frost after a long wet period, since the absorbed water in the paint layer will freeze.

Dust effects, which can lead to pitting corrosion of aluminium alloys and other materials, must also be taken into account if necessary. Model tests can be devised without difficulty if the chemical composition of the dust, and other factors which affect it such as washing of the surface by rain, are known and understood. Reliable and meaningful data can be obtained from such tests if they are sufficiently long [20].

As was mentioned earlier, simplified models of atmospheric corrosion can almost always be devised in the laboratory. This can only follow a basic analysis of all quantitative and qualitative factors affecting the action and intensity of the corrosion in the natural state, however. There must be a qualitative determination of the 'dominant combination of factors', and it is preferable that this should be quantified, if possible.

It seems obvious that there is no universally applicable model test method. Each point in the test program needs special consideration, and adaptation of the model test if necessary. The more precise the question which is posed, the more exact can be the model test, and the more informative the results.

One further point should be mentioned. A correctly-designed model test causes no acceleration of the corrosion process, and will not yield results over short periods. There are advantages in model testing compared with natural testing, however. The effective parameters are firmly controlled, and no unforeseen accidental natural effect can upset the test. Assessment is possible throughout the experiment in the laboratory, so that the corrosion can be evaluated far more often, without difficulty, and several methods can be used to do it. Reproducibility is better, and it becomes easier to recognize the mechanism and kinetics of the corrosion process. This is particularly useful in tests designed as part of investigative or development programs. If these ideas are taken into account, it becomes possible to obtain, relatively quickly, valuable results by analysing the initial part of the corrosion kinetics curve, etc.

6.4.2   Short-term corrosion tests (accelerated tests)

The generally slow progress of atmospheric corrosion means that, in many cases, there cannot be rapid replies to questions on the suitabilities of different protective measures, or their protective value (i.e. their expected lifetime). Natural tests, and often model tests as well, last months or years. Because of this, laboratory test methods have been sought for some years which would allow technically useful results to be obtained from short test durations.

The basic principle of such test methods appears plausible at first sight: if the dominant combination of factors is known, a 'strengthening' of their action should lead to acceleration of the corrosion. This is undoubtedly true in theory, but difficult to put into practice.

Closer investigation of the 'accelerated' tests which are currently most used (salt spray, condensation apparatus, artificial industrial atmospheres, SO_2 test, etc.) shows that they are comparable to conditions in natural atmospheric corrosion only to a limited extent, or not at all. The results of such tests should therefore be seen simply as very approximate first information, which is scarcely sufficient on which to base important decisions or conclusions. The reason for this failure is undoubtedly the naive method of corrosion acceleration; such acceleration can only be successful if it does not lead to changes in the corrosion mechanism.

The previous methods do not meet this criterion, as can be shown quite easily by simple comparison. If the corrosion products formed on metals in salt-laden marine atmospheres are examined chemically and crystallographically, and compared with those formed during salt spray tests, differences appear immediately. In the salt spray test, much more chloride is found on the surface, and the form of its binding and its crystallographic composition are very different from natural products. There is more amorphous material in products from the salt spray test.

Different corrosion products cannot be produced in the same corrosion mechanism, and so the path of the corrosion in the rapid test must differ from that in nature. Though, for example, the typical ratio of corrosion rates of iron and zinc in natural atmospheres is between 1 and 20 to 1, it can easily happen that in an accelerated test both metals are found to corrode almost equally rapidly. Similarly, it often happens that incorrect ideas on relative corrosion resistances of materials are engendered by such inappropriate tests. For example, it was thought for many years that cadmium was a generally more resistant coating material than was zinc. This was the result of a wrongly-interpreted rapid test in a salt spray apparatus, which gave results which hardly accorded with natural corrosion experiments. For a long time, many mechanical and electrical engineers were convinced that most aluminium alloys were considerably endangered by their low corrosion resistance. These alloys had been found to become covered quite rapidly by an ugly layer of voluminous corrosion products (like rust on iron) during testing in a salt spray cabinet; however, this occurred only in the rarest cases under natural conditions.

Tests which investigate the behaviour of materials and coatings in industrial atmospheres can lead to similar errors. It is certain that the corrosion is promoted by combined action of high humidity and the presence of SO_2. However, exaggerated enhancement of the SO_2 action causes such distortion of the corrosion process that the results are worthless. The different reactions of sulphur dioxide during atmospheric corrosion were dealt with comprehensively in Chapters 3.5.2 and 3.2. If an amount of sulphur dioxide is added to the test chamber, to accelerate the corrosion, which is many orders of magnitude greater than atmospheric SO_2 concentrations, the danger of distortion of the corrosion mechanism cannot be disregarded. The first step in SO_2 conversion is oxidation to sulphate ion, even at the highest concentrations occurring in industrial regions. There are limits to this, however. If the SO_2 concentration becomes too high, sulphite and sulphide are found in the corrosion products instead of sulphate. This obviously indicates invalidation of the original corrosion mechanism. Another effect of over-exaggerated SO_2 addition is the formation of normal salts instead of basic salts.

As was discussed in Chapter 3.7.2, the acceleration by raising the temperature is not generally valid, either.

The following points remain to be clarified:

Is it generally possible to accelerate the corrosion process without changing the corrosion mechanism?

If yes, on what basis can the acceleration be achieved?

What limits of use apply to accelerated corrosion testing?

Can generally applicable tests of this type be developed?

It may be concluded from an exhaustive analysis of the current knowledge of atmospheric corrosion processes that it must be possible to have both acceleration and an unchanged mechanism. As was discussed in more detail in Chapter 3, atmospheric corrosion is a discontinuous process which alternately proceeds and stands still. The reaction rates during the individual corrosion periods can thus be defined as functions of climatic factors (or factors associated with them).

The overall corrosion is related above all to the total of the 'active' intervals. It should therefore be possible to increase the frequency or length of these periods in comparison with those occurring naturally, or even to allow the corrosion to proceed uninterrupted. This idea is certainly tempting, but is unfortunately not without problems. As long as there is no effect by the 'dormant' periods on the corrosion process, this idea is acceptable, but examples in which the drying-out period play a significant role are not rare. Solid corrosion products may change their properties during these periods (e.g. by crystallization); organic protective coatings lose a part of the water they take up during the active periods, which undoubtedly affects the rate of penetration of water to the metal-coating interface. If the drying-out periods are simply ignored, phenomena can develop due to the accumulation of adsorbed water which are not observed normally (e.g. blister formation under the coating). The acceleration of the corrosion process by lengthening the total time of active corrosion is thus possible, but only within certain limits which must be determined by prior experiment.

The second method of accelerating the corrosion process without upsetting the mechanism should be utilization of the dependence of atmospheric corrosion kinetics on climatic parameters which can be adjusted in the laboratory. The relationships for unprotected metals were described in Chapter 3.8. Three properties of the surface electrolyte play roles: the activity of the water (which depends on the atmospheric humidity or on precipitation), the activity of the stimulating ions (which depends on the degree of atmospheric pollution), and the temperature. There are possibilities of accelerating the corrosion by enhancing these atmospheric parameters, singly or in combination.

Enhancement of the water activity is easily obtained. Water can be allowed to condense in a condensation chamber, or the humidity can be adjusted to 100% to allow a short period of condensation (by varying the temperature). If the danger of exaggerated action, which has been mentioned several times, is kept in mind, there are no special problems with this type of acceleration.

It is difficult to find limiting values for acceleration by increased stimulator action. The reaction paths by which stimulators accelerate corrosion reactions in nature are complicated, and individual steps in the conversion reactions cannot be neglected. If, for example, the SO_2 action is to be enhanced, the limiting value for acceleration is probably related to the quantitative rapid conversion of the sulphur dioxide on the surface to sulphate. As long as the amount of SO_2 is small enough, this condition is met, and the danger of distortion of the mechanism is small. If this SO_2 concentration is exceeded, the sulphur dioxide begins to react as a solution species (i.e. HSO_3^-), and this leads to corrosion reactions (and corrosion products) which differ markedly from natural ones. The limiting value can be determined experimentally [20]; it lies far below the concentrations used in many standard tests (at about 10^{-3} to 10^{-2} volume per cent).

Enhanced dosing with chloride is beset by similar difficulties. If too much salt solution is added, not all of the chloride participates in the corrosion reactions [21]. There is a large excess of chloride in the corrosion products, which not only alters the corrosion product composition (in place of well-protecting basic salts in presence of large excess of hydroxide, soluble salts form, for example), but also causes acceleration of the corrosion because of their capacity to absorb water (hygroscopicity). During the drying periods which must be included at frequent intervals in the test program (as was discussed earlier), the water is not sufficiently removed.

The accelerating action of high temperatures is often over-emphasized. Test methods usually use temperatures of 35 to 40 °C, and this can be attributed above all to superstition. Though not completely negligible, temperature is a minor parameter in accelerated testing of corrosion of metal surfaces (though it is highly significant in testing of organic coatings). Atmospheric corrosion under natural conditions normally occurs between 0 and 30 °C. It was shown in Chapter 3 how rusting of ferrous metals is the only corrosion reaction which proceeds more rapidly at elevated temperatures.

Significant increases in temperature (in comparison to those found during active corrosion periods in nature) bring dangers with them. Relative corrosion rates of different metals change: the rate on iron rises, while those of most non-ferrous metals do not. Higher temperatures at the same relative humidities and unchanged stimulator contents imply significant increases in absolute water vapour content and so in the ratio of the two important parameters i.e. the water vapour to stimulator activity ratio changes at the expense of the stimulator. The stimulator cannot then act as before, since it is, so to say, 'suppressed' by the over-enhanced water activity. As a result, the corrosion may proceed more slowly than at lower temperatures. Analyses of corrosion products also suggest that this method does not give proper acceleration. Though there are relatively large amounts of stimulator in the test atmosphere, only very small amounts of their conversion products are found in corrosion products [22]. It is perhaps more suitable to produce the acceleration by holding the ratio of water and stimulator activities constant in the test atmosphere.

Similar ideas apply in the derivation of accelerated testing methods for complex systems of protectively coated metals. It is difficult to devise useful accelerated test programs if there is no exact knowledge of the mechanisms of destruction of these systems. The problems are particularly great for organic coatings, since the properties of such coatings are various, and their protection arises for different reasons (see Chapter 5.6). Compatibility of the coating with water (i.e. uptake without swelling) must be taken into account, as must aging phenomena caused by photochemically and thermally accelerated processes.

There is very little known about the mechanisms of weathering action on organic coatings. Every binder-pigment system and every multi-layer coating of paint poses new problems because of the huge number of factors to be considered (e.g. water uptake and permeability, swelling, permeability of aggressive species, photochemical sensitivity, and adherence to the base metal and between layers), and all of these must be considered in developing an accelerated test. Only very general principles can thus be used.

Accelerated photochemical aging of the coating is due to action of energetic UV radiation on water-containing coatings. The simulated radiation which is used must not have too high a component of short-wave light, which would not correspond to that found in the solar spectrum. Use of irradiation and relative humidity as parameters provides data on corrosion of the coating which can be very usefully analysed.

The influence of frost temperatures on organic coatings should not be neglected, and must be included in the test program. (There is no effect of frost, as such, on corrosion of metals.)

Such considerations form the basis of the different test cycles in which drying-out periods alternate with periods of intensive action of water (e.g. in a condensation apparatus) and periods of combined action of water and radiation, with introduction of periods of the action of frost on damp surfaces. It is senseless, however, to repeat the different parts of the cycle in rapid succession. This does not happen in nature, and is not desirable from the point of view of corrosion acceleration. An organic coating takes up water relatively slowly, and does not lose it rapidly. Water cannot, therefore, penetrate deeply enough into the coating during rapid cycling of the test conditions, and many phenomena which would occur under natural conditions disappear. This distorts the corrosion mechanism, and invalidates the accelerated test. Periods of action of constantly enhanced combinations of factors which last several hours, which would simulate the typical daily corrosion periods in most aggressive seasons, are to be recommended.

The well-developed processes for measurement of the value of bright chromium coatings occupy a special place in the group of short-term tests. They permit reliable ranking of different coatings which agrees qualitatively with their behaviour in operation (i.e. system A is better than system B, etc.). They are the popular CASS (Copper–Acid–Salt Spray), developed from the standard salt spray experiment, and the even faster Corrodkote method. Though these methods are based on empirical principles (simulation of the composition of smut deposited on chromed automobile trim in a large city in USA), they are outstanding at 'modelling' the conditions for the appearance of pitting corrosion of the intermediate copper and nickel layers. It is well-known that nickel is attacked by pitting corrosion; this local disruption of passivity is caused by high chloride concentrations and certain redox potentials. The chloride concentration is governed by the solution concentration in the CASS test, and is fixed exactly by the amount of chloride in the kaolin suspension in the Corrodkote test. Cupric ions are added to the test solution in sufficient quantity to set up the $Cu^{2+} + e \rightarrow Cu^+$ redox system to produce a sufficiently positive constant redox potential. The electrolyte pH is adjusted to the optimum value using acetic acid in the CASS test. From experience, the Corrodkote test provides reliable results. This applies a suspension of kaolin, which is inert in corrosion, in a solution containing Cl^-, NO_3^- and Cu^{2+} as active components to the surface under test, following a set scheme. The simple exposure conditions (in a humidity cabinet) help to improve the reproducibility [23].

Instead of regulating the redox potential by adding oxidizing ions, the same result may be obtained by anodic polarization. This is the basis of EC (Electric Current) tests [24]. It is possible to maintain a potential which favours pitting corrosion of nickel by potentiostatic or galvanostatic polarization, and to produce, in a solution containing only Cl^- ions as the aggressive constituent, similar corrosion phenomena to those found during deterioration of bright chromium layers on vehicles on the road. These methods have come into use, in particular, because of their extreme rapidity of performance. One modified form of these methods is particularly promising. It permits determination of the speed of corrosion after only one-minute cycles by development of colours on the surface at points where pitting is developing [25]. Colorimetric determination of the metal which has gone into solution is convenient, but gives no insight into the number or distribution of sites of pitting.

It seems, therefore, that these empirically-developed test methods are good examples of rapid test methods.

Another important question in rapid corrosion testing is: For how long does a test give worthwhile results, or how long must it last to do so? The answer to this is embodied in the question; the test must last long enough for recognition of sensible differences between specimens, to give qualitative results. Another possible answer to this question is based on the following ideas: every atmospheric corrosion process has, during the course of the practical life of the protected or unprotected material, a period with a fairly constant corrosion rate. Until this is reached, the corrosion proceeds faster (on unprotected materials) or slower (especially metals protected by an organic coating). It may increase again later (e.g. after the protective action of the coating is lost). The most important part of the life of the material is during this stationary corrosion period. The rapid test should therefore be allowed to run long enough for the stationary rate to be clearly attained, and this result can then be compared with that from other systems to provide more definite results [20, 21]. This idea is shown graphically in Figure 58.

152

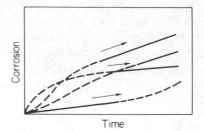

Figure 58. Comparison of corrosion
rates from laboratory tests.

If it is possible to find relationships between stationary corrosion rates in natural atmospheres and in rapid tests, it is possible to calculate so-called coefficients of acceleration. These have been discussed widely, and are usually regarded as unrealistic [20, 21].

Just as in natural and model testing, more valuable results are obtained from short-term tests if more appropriate, more complex, and more quantitative methods of evaluation are chosen.

Satisfactory acceleration of atmospheric corrosion in the laboratory is still not a completely solved problem. One thing is certain: no generally applicable rapid test can be devised. The purpose of each test and each test system requires special consideration, and appropriate test conditions and assessment procedures. In this respect, the standard test methods (salt spray, SO_2 test, etc.) cannot be rejected completely, since they may form parts of more complex, but more suitable, test programs. A warning against the overstatement of their general applicability should be sounded, however, since their use by less-experienced investigators may lead to false conclusions.

6.4.3 Indirect tests

In contrast to model and rapid tests, 'indirect' tests are not performed in natural atmospheres. The properties of the metal (or metal-coating system) which are most important for corrosion are measured, and the measurements are assessed with a view to the probable corrosion behaviour of this material in a particular natural atmosphere.

Such predictive tests can often be performed very quickly. Reliable results should only be expected, however, if specific relationships between the measured property and the corrosion behaviour are known. Unfortunately, the knowledge of such relationships is so far very limited, so that obvious methods can often not be used as reliable test techniques.

Some valid examples do exist, however. It is senseless to investigate the corrosion of zinc coatings on steel by using model or rapid tests, unless a special effect is being examined. Normal weathering resistance (and hence life) of such coatings is unequivocally related to the coating weight (or thickness) [26]. Effects due to process technology (i.e. different methods of producing the coating) play a minor role here, as long as the protective coating contains mainly zinc. Pores in the coating are similarly unimportant. Thus, if the rate of corrosion of zinc in a typical, or better still in an actual, atmosphere is known, the life of the coating may be calculated quite reliably. The most important property of the coating is its thickness, and indirect evidence on the corrosion protection value of the coating is given by measuring this. The layer thickness measurements may be complemented by other data; thus, on thermally sprayed coatings, useful extra measurements may be made of the coating weight, uniformity of layer thickness, and layer adhesion. These are all further indirect gauges of the likely corrosion properties.

Similarly, aluminium coatings need not be investigated by direct testing. The extremely stable passivity of these coatings, which disappears locally only under intensive action of salts, means that one must wait years or months for useful results from laboratory studies. The protective action is due mainly to coating passivity and sealing of small pores by corrosion products, and sufficient data for estimation of corrosion resistance can be obtained from measurements of coating thickness and pore properties (e.g. by using chemical or electrolytic reactions of coloured indicators). However, since it is difficult to differentiate which pores will be sealed at the onset of corrosion, and so will be harmless, and which are too large to be sealed, it is desirable to carry out pore measurements only after model or rapid testing to the point where pores should be sealed. Examples of methods for this are mild salt spray testing or SO_2 action in a condensation chamber.

Measurements of thickness, pores, and adhesion also characterize other coatings sufficiently for direct tests to be unnecessary. This is true of lead, cadmium, and other coatings which are eroded evenly in the atmosphere.

However, if the corrosion of the coating is of the pitting type, direct tests cannot be avoided. A typical example is decorative chromium coatings on copper–nickel underlayers. The accelerated methods discussed in Chapter 6.4.2 (Corrodkote, CASS, EC) give, to date, faster and more useful characterization of these coating systems than do indirect assessments. Typical indirect measurements here might involve a combination of metallography and determination of crack and pore shapes and chromium coating density, or electrochemical characterization of the individual layers by potential or polarization measurements, etc. Obviously, the relationships which determine the corrosion resistance of these coatings are too complex (in relation to the simple ones for zinc) to allow complete determination of the resistance to corrosion by indirect measurements.

Up to this point, the effect of internal stresses in electrolytic coatings on their corrosion resistance has not been taken into account. The measurement and evaluation of these stresses and their effects are undoubtedly measurements which permit predictions of the corrosion behaviour of the coating [27].

As the comparison of the efficiency of indirect measurement of corrosion resistance in simple systems like zinc coatings and complex bright chromium coatings has shown, simpler protective systems are far easier to assess by indirect methods than are complicated ones. The reason is undoubtedly the lack of understanding of the exact protection mechanism in the multi-layer system, which prevents the measurement of exactly characteristic properties of the system.

This is also true for organic coatings or paint layers, which have widely varying modes of action. The formation of series of layers, the influence of surface pre-treatment, the properties of the different layers, their compatibility under different external stimuli, aging phenomena, etc., all combine to make it impossible to differentiate clearly the properties which are critical for long-term stability of corrosion protection, and thus the properties which should be measured. Layer thickness measurements and pore determination are only parts of the information needed, and must be supplemented by special model or accelerated direct testing.

Other parameters can also be determined to allow indirect evaluation of corrosion behaviour. For example, if certain environmental parameters are known to affect the corrosion or stability of a material, it is sufficient to measure these. An example of this is continuous measurement of climatic factors in a warehouse, from which the danger of water condensation, and hence the probability of corrosion, may be assessed.

There are many atmospheric corrosion effects which are produced on a surface by a specific factor; indirect measurements can also be usefully applied in these cases.

7
Technical and Scientific Considerations in Atmospheric Corrosion Protection

The concept of 'atmospheric corrosion' embraces a large number of practical problems which are faced by technologists. Their satisfactory solution requires a complex analysis of all the factors which influence the choice of a method of corrosion protection. These include not only descriptions of corrosion effects (and aggressiveness of the atmosphere), but also questions relating to the scientific and technical requirements of the products to be protected, the process technology (such as advantages or difficulties in using a particular protection system), etc.

The individual steps in this analysis may be loosely broken down as follows:

Determination of the scientific and technical aims of the protection.

Characterization of the aggressiveness of the atmosphere.

Choice of the appropriate type of protection (short- or long-term).

Concentration of possible methods down to choice of the optimum system, in consultation with the manufacturer, testing technician, etc.

7.1 Determination of scientific and technical aims

The technical requirements for satisfactory corrosion protection depend on the purpose of the protection. The following possibilities are available:

Maintenance of the mechanical properties of a material, which are the reason for its use in a specific application (e.g. steel structures, motor vehicles, etc.).

Guarantee of defined surface properties, such as conductivity, surface state (sharpness, polish), optical properties, etc. There should be a basic distinction drawn in this group between long-lasting functional protection and necessarily temporary protective measures (e.g. during storage or transport).

Maintenance of a desired aesthetic appearance of the product.

The first two classes may be defined in terms of the desirability of corrosion product formation. If a simple question of prevention of metal erosion by corrosion is involved, corrosion products are usually acceptable. In the second class, where the physical surface properties and the changes in them caused by corrosion define the worth of the protective method, the formation of corrosion products is generally undesirable.

The classes overlap in particular cases. Aesthetic features predominate in corrosion protection of an automobile body, during the economic lifetime of the vehicle, but the corrosion must not impair the strength of the steel, either. Many metal surfaces must be protected with respect to both of these criteria.

The economic aim of all corrosion protection, and so of atmospheric corrosion protection, is the achievement at minimum cost of all corrosion protection measures throughout the whole satisfactory life of the product. A consideration of the expected lifetime is very important in the choice of practical protective measures. The total costs of corrosion protection include not only the costs of initial production but also the costs of maintenance needed to keep the product to its specified performance throughout its life. Laying aside the fact that repair may be associated with considerable difficulties, the cost of upkeep of very long-lived technical equipment, such as masts or bridges, in aggressive atmospheres is very high. It may be necessary to build auxiliary scaffolding, and to pay high wages; the work force required for such tasks, which are often dangerous, is frequently difficult to obtain in industrial regions.

If a rusted steel structure is to be given a new coat of paint without the old one being removed, the new coat is not only more expensive, but its protective value is much less than one applied during manufacture or erection. The reasons for this are clear: the renewal must be done in the open air, the work can be done only in warm, dry weather, the removal of the rust and the old paint is difficult and very costly, inaccessible parts pose problems, etc.

Indirect costs should also be taken into account. For example, repainting of electrical supply pylons often necessitates the interruption of the power supply, which is hardly economical.

From the point of view of safety, it is often wise to use very expensive protective measures on products whose function and reliability may be especially endangered by atmospheric corrosion. This applies to regulatory apparatus, for example, whose failure could cause considerable damage.

It is valuable to know what life expectancy and reliability of the protection may be expected on a particular product. A customer is often prepared to pay a raised price for guaranteed protection.

Once the technical aim of the protection has been clearly established, and the desired life of the product is fixed, alternative methods of protection against corrosion can be examined. Further development work should then lead to choice of the optimum protection method. The rest of this chapter discusses the ideas and methods of implementation of this decision process.

7.2 Consideration of the aggressiveness of the atmosphere

7.2.1 Determination of corrosion conditions on long-lived stationary structures

For some stationary installations, the environment is known from the outset, and it is therefore possible to determine definite atmospheric corrosion conditions. This is worthwhile if it can be done, but to date such analyses are performed only rarely (in contrast to the protection of buried structures, such as pipelines, where only rarely are there no preliminary investigation of soil conditions, etc., with respect to likely corrosiveness).

The protection against corrosion of bridges, masts, television towers and other, usually very expensive, equipment which is required to last a long time should always be chosen after exhaustive evaluation of the aggressiveness of the atmosphere at the site. The methods involved in doing this are not discussed further here; they can be derived from the discussion of climatic and chemical measurements in natural corrosion testing (Chapter 6.3.3) and from the information in Chapters 3 and 4 on the long-term courses of atmospheric corrosion, and principles for evolution of different protective measures. Again, it should be stressed that these measures cannot be based on general data on average values of climatic factors. The most exhaustive possible analyses of individual measurements are required, to give exact information on decisive parameters, such as the duration of active corrosion periods, or stimulator contents in the atmosphere. It is usual to supplement such studies by corrosion testing at the site, and this often produces surprising results; atmospheres regarded as relatively harmless may often be very aggressive, though the source of the stimulator species may be some distance away.

7.2.2 Atmospheric corrosion conditions for general products

In contrast to the rather specialized cases described in Chapter 7.2.1, where there is exact definition of corrosive conditions in the atmosphere, there are great difficulties associated with the consideration of atmospheric corrosion effects on technical products which are produced in large numbers and used in widely varying atmospheric conditions.

A good example of this is motor vehicles. These must have a pre-determined working life, and can be exposed to very different atmospheric action during this time. An automobile must last reliably for 200,000 to 300,000 km under arctic or tropical conditions without having its usefulness impaired by specific corrosive environments, such as salt-rich atmospheres or those containing industrial waste gases. The same applies to railway vehicles, ships, aircraft, and electrical, industrial and household equipment, etc.

It is not possible to apply different protection to an automobile, which must function anywhere in the world, to deal with tropical or industrial or any other special corrosive atmospheres, and the same applies to electric motors, recorders, and many other technical products. However, it is desirable to take into account a specific corrosive environment if this is able to be defined, and to choose appropriate protective measures.

A generally applicable system of classification of corrosive natures of different indoor and outdoor atmospheres can help here. If the various atmospheres in which metals are used in technical applications are classified with respect to their corrosive properties, so that the basic combinations of dominant factors which have been outlined in earlier chapters can be evaluated, it appears possible to define four or five categories of aggressiveness.

The first two categories include indoor atmospheres in which there is such complete control by walls, roofing or heating or air conditioning that corrosion-favouring conditions arise rarely, or not at all.

The more aggressive two or three categories involve typical outdoor atmospheres in which weather conditions often permit longer action of corrosion-promoting effects. Differences arise chiefly in the intensity of action of stimulator species (salt mist, industrial pollutants). Pure rural and forest atmospheres represent the third category, while urban, industrial and marine atmospheres are included in the fourth.

The fifth, most dangerous, category is reserved for special cases which cannot fit into the other four. These might be caused by the specific conditions around a chemical factory or some similar environment, where there are extremely high pollutant concentrations or where the surface comes into contact with attacking solutions.

A classification scheme such as this is founded on two ideas. For non-aggressive atmospheres, such as indoors, the danger of corrosion product formation is of the greatest importance. The probability of the occurrence of corrosion-favouring conditions i.e. of surface electrolyte formation by condensation, is assessed. This assessment is usually expressed not as a corrosion rate (e.g. in terms of material removal), but rather as a degree of danger of the formation of undesired corrosion products. Store-rooms, work spaces, etc., are typically assessed in these terms.

The analysis of climatic data can be adjusted for this purpose without difficulty. It must, however, be accumulated at short intervals (and most preferably, continuously) over long periods in typical weather conditions. The inhomogeneity of the atmosphere in the room should also be considered. Near to windows and doors, especially if they are open, or if the indoor and outdoor atmospheres mix during movement of goods, air currents are formed which may lead to attainment of the dew point at the metal surface, and thus to increased danger of corrosion. The action of the product as a heat sink should also be considered. An object which is moved from a cold atmosphere into a warm room becomes covered with condensation, and this condensation is more marked and more long-lived the heavier the object (and so the greater its heat capacity). Hygroscopic species on the surface can also increase the danger of corrosion, as was discussed earlier.

In spite of these additional factors, the corrosion danger can be satisfactorily deduced by analysis of suitable climatic measurements, and the following classification may be proposed: If corrosion-favouring conditions never, or rarely and then for only short periods, occur during the year, that indoor atmosphere may be designated as free from danger of corrosion. Fully air-conditioned, and most cases of heated, rooms have this type of atmosphere (except around doors and windows). The limiting duration of surface electrolyte presence in such atmospheres is approximately $0 \cdot 1\%$ of the year (i.e. about 9 hours), assuming that there is no stimulator present on the surface which is endangered by corrosion. If this limit value is exceeded, a certain danger of corrosion must be expected, and the atmosphere then falls in the second group.

From a practical point of view, this means, for example, that no temporary protection of surfaces endangered by the corrosion is necessary for those atmospheres in the first category, whereas it is desirable for the second group. An exhaustive analysis of the climate of the indoor space, especially if all additional factors are considered, frequently has surprising results (and uncovers unexpected causes of corrosion), and this helps in finding proper protection methods. Simple temporary protective measures may solve this sort of problem; a pre-conditioning of the air often helps to minimize the danger of corrosion.

There are some indoor atmospheres in which there are frequent long periods of surface electrolyte presence, and classification according to probability of corrosion must be abandoned here. Examples are atmospheres with high impurity contents, such as in pickling or galvanizing plants, or spaces with high water vapour content, such as laundries or indoor swimming pools. Such cases must be included in one of the next categories.

If conditions favour atmospheric corrosion for significant periods of the year (with a total of perhaps over 10 % of the year), the probability classification is no longer useful, and the aggressiveness of the atmosphere should now be described in terms of a corrosion rate. Again, the comprehensive discussion in Chapter 3 of the causes of long-term atmospheric corrosion is helpful here. The long-term corrosion is the sum of a number of individual corrosion periods, which each have individual lengths and individual average corrosion rates. The duration of these periods (and hence their sum) can be deduced reliably from meteorological data for the majority of cases, and the corrosion rates of the different metals can be estimated approximately. Data for most climatic zones suggests that the duration of wetting is 10 to 60 % of the year, though there are some exceptions; smaller values are often found in desert regions, while some tropical forest areas and coastal regions may have higher values. In any case, there is a possibility of breaking the corrosion down into short periods as a basis for classification of the atmosphere, and of setting arbitrary limits.

If, however, the classification is to be used only for a certain zone, limited by geography and climate, for example, a European country (or perhaps the whole of middle and western Europe), this classification scheme is inappropriate. Most technical products will then be used in comparable climates which will not be characterized by widely differing corrosion periods. There are exceptions to this, though they are rare; for example, mountain regions, where wetting periods occur more often and last longer. Over most of Europe, though, the relative corrosion periods cover 30 to 50 % of the year, so that exact differentiation cannot be obtained from this scheme. (Inland USA invariably has a smaller percentage of wetting during the year than this.)

A knowledge of the stimulating action of atmospheric impurities can help in resolving this problem. A lower limit can be set on this action, at which there is still no marked promotion of corrosion, and this can be based on theory (see Chapter 3). If this limit value of atmospheric pollution is not reached, the corrosion is significantly slower (Category 3) than if it is exceeded (Category 4). Empirically-determined values (gathered from long-term investigations of corrosion in natural atmospheres) are already available to serve as bases for standardization here.

Tables 14 and 16 (see Chapter 7.5.1) contain a classification system which is based on these criteria.

This system also covers the special cases which fall outside the usual limits. These special cases must always be considered individually.

Specifications for surface protection methods can be drafted using such a classification system, and can be used as the basis of a similarly derived quality classification. The ISO recommendations on galvanized coatings have four quality grades, for example, based on a system like this.

There is a clear relationship between quality specifications and the type of test required. The 'strength' of the test can be suitably adjusted to the classification system. Such a classification scheme can clearly be applied to a particular product which has specific economic lives in different applications, so that it is possible to set quality grades for protection of metal structures, vehicles, electrical apparatus, or other technical equipment, which may function indoors or outdoors, and which must deliver a specified performance for a pre-determined period without impairment of its usefulness or reliability by atmospheric corrosion.

7.3   Choice and implementation of corrosion protection

Once a particular group of protection methods have been selected for use on a particular product, using the principles outlined above, the implementation of the protection begins. In this, the corrosion engineer must take into account all the manufacturing and process details which may affect the quality of the protection. The adage, 'Corrosion protection begins on the drawing board', is not too fanciful; in fact, it might be extended even further. Even before the final design is conceived, there must be simultaneous basic design of the product and planning of corrosion protection.

The corrosion engineer in the development team must take account of the following points, in particular:

The general conditions under which the product will be used.

Possible establishment of special surface microclimates due to the function of the product and the arrangement of its working parts.

Exclusion of corrosion-promoting bimetallic couples, which could produce electrolytic corrosion.

Possible effects on corrosion behaviour of certain manufacturing processes.

Optimization of the structural aspects of the product with respect to the chosen corrosion protection, to make the protection technology as simple as possible; there should be no inaccessible places, and there should be suitable finishing of the surface.

The possibility of repair of the corrosion protection, should it be needed (as it usually is on long-lived equipment).

7.3.1 General principles of construction for proper corrosion protection

The general data on corrosion rates and economic lives of different metals, alloys, and protective coatings are of use only if important principles are observed in their use. Practical experience and results of experiments show that a material or coating, in the same atmospheric environment and under identical exposure conditions, can corrode at different rates [1]. Very many influences are at play here, which must all be considered, at least qualitatively, when the corrosion protection is being planned (even on the drawing board) and implemented.

The arrangement of the parts of the product, their mutual orientations, and functional factors are one group of extra influences which can cause marked deviations from corrosion rates measured in simple tests.

The very first point to be considered is whether certain types of finish do not affect the microclimate at the surface. If there is heat exchange at the surface, whether it is expected or not, this possibility cannot be ignored. A cold water pipe in a damp, warm atmosphere is always wet because the dewpoint is always exceeded at its surface, and so corrosion can proceed uninterrupted, and may be far more dangerous than would be assumed. If equipment is radiating heat as it works, there will be markedly decreased corrosion of surfaces held at temperature above the dewpoint by this action. The change of temperature, and so of relative humidity, is sometimes not limited to the heat-exchanging surface, but affects adjacent surfaces as well.

Special localized surface microclimates arise from the actual structure of the product. Rain or condensed water should flow off all parts of the product, or be evaporated rapidly. There are important effects played by crevices, grooves, cavities, and the orientation of surfaces with respect to the ground. It should be noted that simply-arranged surfaces not only provide advantages in surface protection technology (see later) but also reduce the danger of corrosion which is posed by specific localized changes in corrosion conditions. Modern construction components, which use few L-, H-, U- or similar sections, but replace them with closed sections, also favour easier corrosion protection. For equivalent static properties, closed sections are not only simpler and more easily accessible for surface protection, but have significantly smaller surface areas which must be protected [1].

It is possible to design for a uniform corrosion attack, by suitable choice of components, and so make the differential protection which is otherwise required superfluous. Corrosion damage can rarely be completely prevented, however; horizontal surfaces always corrode less on the surface facing away from the ground than on the surface facing towards it, and specific actions of microclimates near the ground can rarely be excluded.

The possibility of removal of water from individual surfaces by flowing off of rain or condensed water and by rapid drying by air currents should be planned for in the design. A trivial example involves use of L- or U-sections in configurations in which there can be no collection of water in the channel; this is, of course, the usual practice already in construction methods [2]. However, protection of closed-profile sections in structure which have been built properly (from a corrosion viewpoint) still poses some complicated problems. It is often difficult to obtain an air-tight seal on hollow components, and a cavity like this which comes into restricted contact with the external atmosphere can pose special corrosion dangers. Water vapour can enter the cavity easily during damp, warm periods, and then condense to water inside on cooling. This water is not removed, and so produces a specific corrosion-favouring microclimate in the cavity. This phenomenon of 'inside-out' corrosion is not rare, as is evidenced by its frequent appearance in corrosion damage to motor vehicles (especially on sills and doors). Creation of drainage holes to allow removal of the water is a solution, though only in part, to this problem.

A locally increased corrosion rate is often found around crevices and corners of structures. This is again related to, among other factors, delayed drying-out of surface water. Water which enters crevices dries out only very slowly, especially if accumulated dust or existing corrosion products are influencing the humidity.

The structure of the product should, if possible, permit uniform aeration of all surfaces. Surfaces lying very close together and aerodynamic 'shadows' should be avoided.

7.3.2 Contact between different materials

The peculiarity of atmospheric corrosion, that the very limited amounts of electrolytes act only temporarily on the surface, reduces the danger of corrosion acceleration by electrolytic short-circuit macro-cells, but they cannot be ignored completely. Though the longest distance over which the electrolytic action is likely to reach is at most a few millimeters, and though the covering layers which form are able in some cases to inhibit this corrosion due to their resistance, there are some cases in which these short-circuit cells can be dangerous.

It is steadily becoming more unusual to find obvious cases of typical corrosion damage in the atmosphere caused by electrolytic action due to the incorrect combination of materials. It seems as if builders keep this area of corrosion science in mind, and follow the rules of thumb or tables of acceptability of contact of different metals.

There have been frequent attempts to classify the different metals and alloys into groups (e.g. Table 12), such that direct contact between metals from within a group should not lead to danger of electrolytic acceleration of the corrosion [1]. Contact between a metal and one from an adjacent group should produce only mild corrosion, and need be excluded only in very aggressive environments (e.g. directly beside the sea). Contact between metals from more widely separated groups is more dangerous, and can lead to accelerated corrosion of the less noble metal even in quite mild atmospheres. These rules of thumb are adequate

Table 12. Division of the metals into 'compatibility groups'

Group	Metal
1	Magnesium alloys Zinc
2	Aluminium and aluminium alloys
3	Iron, steel and cast iron (active) Lead–tin solder metal Lead Tin
4	Nickel Nickel alloys (e.g. Inconel) Brass Copper Bronze Copper–nickel alloys Stainless steels (passive) Monel
5	Silver solder metal Silver
6	Gold Platinum

in most cases, especially if they have been based on satisfactorily organized tests under natural conditions.

Different handbooks contain other tables of corrosion protection and compatibility, which provide data on the danger of electrolytic corrosion acceleration from direct contact between different metals (e.g. Table 13) [3].

It should always be remembered in applying this data in practice that if one of the metals which forms the couple is present as a coating (e.g. on steel), there may be effects due to the corrosion-influencing properties of the coating, which normally adheres to its substrate only for a limited period. The coating must accordingly be pore-free and sufficiently thick for it not to be prematurely destroyed, which would then expose the substrate to the corrosion-promoting effects of the couple. It was discussed earlier how the danger of electrolytic corrosion acceleration may be over- or under-estimated when a noble metal is in direct contact with an electronegative metal; the data in Tables 12 and 13 must accordingly be regarded simply as a guide, and where necessary should be supplemented by special testing. The tables should also be extended for those cases of direct contact between different metals which are forbidden, to determine the corrosion danger accurately and to try to find remedies. These may involve the use of metallic coatings, or, if these are also forbidden, an organic protective layer on one or both of the metals which are to be brought

Table 13. Effects of metallic contacts on atmospheric corrosion

Number	Investigated metal	Affected metal										
		1	2	3	4	5	6	7	8	9	10	11
1	Carbon steel, cast iron	—	B	B	B	B	B	A	A	A	B	B
2	Stainless steel	A	—	A	A	A	A	A	A	A	A	A–B
3	Nickel and nickel alloys	A	A	—	A	B	A	A	A	A	A	A–B
4	Chromium	A	A	A	—	A	A	A	A	A	A	A
5	Copper and copper alloys	A	B	B	B	—	A	A	A	A	A	B
6	Aluminium and aluminium alloys	B	B	B	B	C	—	A	A	B	A	C
7	Zinc and zinc alloys	C	C	B	B	C	B	—	B	B	B	C
8	Cadmium	B	B	B	B	B	B	A	—	A	A	B
9	Magnesium and magnesium alloys	C	C	C	C	C	B	B	B	—	C	C
10	Lead, tin and their alloys	A	B	A	B	B	A	A	A	A	—	B
11	Silver, gold, platinum (rhodium, palladium, osmium, iridium)	A	A	A	A	A	A	A	A	A	A	—

Columns 1 to 11 correspond to the description on the row with the same number.

A: There is no effect of contact of the metals on the corrosion of the investigated metal.

B: There is a small effect of contact of the metals on corrosion of the investigated metal, but it is significant only in very aggressive atmospheres.

C: Strong effects in outdoor atmospheres, and also exceptionally, in aggressive indoor climates.

into contact. Insulating washers or gaskets may also be used, though this is used more rarely in practice (and usually on screwed or bolted joints).

The danger of electrolytic corrosion must also be considered particularly carefully where strongly electronegative light metal alloys are used. The two important groups of materials, aluminium and magnesium alloys, will therefore be discussed more closely in this respect.

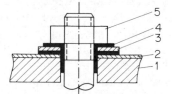

Figure 59. Insulated bolt arrangement.

1. Base metal; 2. Paint layer; 3. Insulating washer; 4. Caulking material; 5. Nut and bolt (noble).

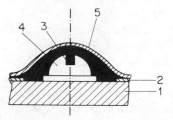

Figure 60. Insulated screw head.

material; 4. Screw head with washer; 5. Paint.

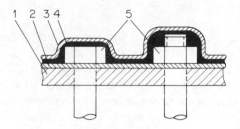

Figure 61. Insulated bolts.
1. Base metal; 2, 3. Paint; 4. Caulking
material; 5. Nut and bolt (noble).

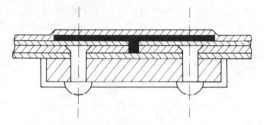

Figure 62. System for insulation of a noble-metal
rivet.

Aluminium and aluminium alloys appear at the base end of the potential series. However, the high stability of the passive layer which forms on pure aluminium and on copper-free single phase alloys causes a marked protection in most atmospheres, so that the greatest danger from atmospheric effects is posed by chloride ions which might be present (e.g. in coastal areas). Since atmospheric corrosion of aluminium is affected little by presence of SO_2 (Chapter 3), normal atmospheres in urban and industrial areas are no more aggressive than unpolluted rural atmospheres, as long as they do not contain chloride-containing dust (e.g. from de-icing salt spread on streets during winter).

Contact between pure aluminium and aluminium alloys, or between the different alloys, is harmless in most cases, but there can be dangerous effects in chloride-containing atmospheres. An example is steel plates which have been plated with pure aluminium and fixed together with Al–Cu–Mg rivets, which has led to problems in the past. There can be similar problems with contact between zinc- and magnesium-containing alloys and more noble (copper-containing) alloys. This problem can be fairly easily solved, however, by choice of chemically similar alloy types with suitable mechanical properties, and by surface treatment of the components (e.g. anodizing or painting) before and/or after erection.

Aluminium–zinc and aluminium–cadmium contacts are always harmless. If zinc is used as a coating on steel, it must be used in sufficient thickness to prevent premature exposure of the substrate. An additional coat of paint causes a significant increase in the life expectancy of the zinc (or cadmium) layer.

Though lead is a noble metal, its contact with aluminium is not dangerous. It is possible that this is due, at least in part, to the surface layer which forms on lead in almost all types of atmosphere, and to the high overpotential (and hence inhibition) of the cathodic reactions involved in the corrosion.

Nickel and copper-free nickel alloys accelerate the destruction of aluminium by contact corrosion only in very aggressive chloride-containing atmospheres. In industrial atmospheres, aluminium-nickel contact is not dangerous. The same is true for aluminium–stainless steel, aluminium–chromium, and aluminium–titanium contacts. The stability of the passivity of chromium, titanium and chromium (nickel) stainless steels and the rapid 'healing' of locally depassivated areas on aluminium are responsible for this compatibility. Bright chromium coatings are equivalent to nickel surfaces in this application, because of their porosity, and so some acceleration of the aluminium corrosion in marine atmospheres should not be excluded.

The corrosion of aluminium by contact with iron (or steel) is accelerated only in chloride-containing atmospheres. This combination should be avoided, however, if the aluminium is being used because of its surface appearance, since there is a danger of its discoloration by 'rusty' water unless the iron has been given a metal (zinc or cadmium) coating or has been painted before erection.

Contact corrosion in the aluminium–magnesium couple affects both metals in aggressive atmospheres. The anodic magnesium undergoes accelerated corrosion, and the alkaline pH produced by the corrosion products and the cathodic reaction at the aluminium cause activation of the aluminium because of the solubility of its passive layer in alkaline solution. This leads to so-called cathodic corrosion, in which aluminium complexes are formed. The higher the potential difference between the aluminium and magnesium alloys, the greater the danger of corrosion of both metals. Thus, if magnesium parts must be joined using aluminium screws, rivets, etc., copper-containing aluminium alloys should be replaced by alloys containing magnesium or zinc.

Metallic contact between copper (or copper alloys) and aluminium should be avoided at all costs. Such couples are trouble-free only in completely inactive atmospheres. Anodizing or painting are of no help in even slightly aggressive atmospheres. If there is no other possibility (e.g. in certain electrical components), the copper must be coated with a sufficiently thick layer of a metal compatible with aluminium (e.g. lead, zinc, or cadmium), or the part must be sealed in an airtight unit.

Magnesium alloys are at the very base end of the potential series of technical metallic materials, and their use in aggressive atmospheres causes problems. The following principles are recommended: magnesium parts should, where possible, be in contact only with base metals. Among possibilities here are zinc alloys or thick zinc coatings on other metals (e.g. steel), thick cadmium coatings, or copper-free aluminium alloys with high magnesium and zinc contents.

Even so, it is still necessary to use further protective measures. The parts should be passivated by chromate treatment, and be painted before and after erection with a good paint system which is impenetrable to water (with a wash primer and chromate primer as the first coats).

Parts produced from zinc alloys should not be exposed to aggressive atmospheres because of the danger of contact corrosion, unless connections to other metals are made by measures to counteract this danger. Zinc- or cadmium-coated contacting metals are recommended, and most aluminium alloys do not pose dangers. Nickel, chromium, stainless steel, etc., are less hazardous than copper-containing alloys, as is the case for aluminium alloys.

Iron (steel) is also a non-noble metal, so that its atmospheric corrosion can be accelerated by contact with noble metals. However, this occurs rarely in practice; only exceptionally are unprotected steel components found in contact with more noble metals under atmospheric conditions.

7.3.3 Joints

Even the points at which different components of a product are joined can have a marked effect on the resistance of the product to atmospheric corrosion. Considerations similar to those mentioned in Chapter 7.3.1 (on general construction principles) also apply to choice of junction types, units and methods. It has already been mentioned that collection places for water or dust, unaerated surfaces, etc., must be avoided. The following points also deserve mention:

Recognition of the danger of crevice corrosion.

Possible decreases in protective coating value at edges and other irregularities.

Correct choice of junction units in bolted or riveted joints with respect to the danger of contact corrosion.

Possible endangering of the aesthetic appearance of protective coatings or of the material.

In the space of this book, there is not space to discuss the theoretical aspects of corrosion acceleration in crevices and grooves. Further data on the significance of the peculiarities of corrosion processes in narrow crevices may be found in the monograph by Rosenfeld [1].

Crevices can raise or lower the corrosion rate. The differences between corrosion processes inside and outside the crevice depend on alteration of properties of the electrolyte. Inhibited oxygen access can cause differential aeration cells, in which the surface inside the crevice acts as an anode and the surface outside the crevice is the cathode. This effect is small, however, because of the limited amount of electrolyte in atmospheric corrosion, unless there are dust or corrosion products in the crevice which increase the duration of electrolyte presence.

Crevice corrosion of steel is especially dangerous in impure atmospheres. The anodic behaviour of the surface inside the crevice leads to accumulation there of stimulator species, and so to intensive acceleration of the corrosion process. Spot-welded parts in motor vehicles are a widespread example of this effect, where no attention has been paid to good surface protection and sealing-off of crevices.

Reduced access of oxygen and accumulation of stimulators can activate passive metals (e.g. aluminium, stainless steel) in crevices, so that associated dangers of rapid corrosion of activated surfaces arise, which can, under some circumstances, be still further promoted by action of an active-passive cell.

Voluminous corrosion products can sometimes act as a wedge to deform or crack apart the joined components.

It is difficult to provide clear data on the effects of crevice dimensions. The highest corrosion of steel seems to occur with crevice widths of about 0·25 mm. Narrower crevices seal themselves fairly rapidly with corrosion products, and so prevent admission of electrolyte. Wider crevices do not inhibit transport processes (Figure 63).

It should also be noted that crevice corrosion can arise from contact between metals and non-metallic materials (both organic and inorganic). Rubber and plastic washers are frequent causes of unexpected corrosion accentuation. Components which can be leached from the non-metal (e.g. sulphur compounds from rubber) can also play a role here.

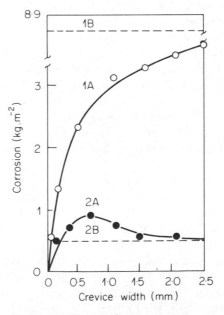

Figure 63. Degree of corrosion after 3 years of steel, as a function of crevice width.
1. Industrial atmosphere; 2. Mild rural atmosphere. A. Inside the crevice; B. Outside the crevice.

Welding has now made the use of multi-material joints, such as those using rivets or screws, almost superfluous, which is a good development from the point of view of corrosion prevention. However, there are still many cases where there is a real danger of crevice corrosion. It is still desirable to work as exactly as possible, to avoid the introduction of crevices during spot welding or due to deformation. It is frequently desirable to use smaller spaces between weld spots or other joints than would be needed for mechanical stability, for this reason.

Good surface protection before and after joining operations is especially essential if steel is involved. There is a large range of priming paints which can be welded without danger to health, and are sufficiently resistant to heat to be destroyed only at weld points. There are advantages in painting surfaces before making bolted joints through them, especially if a further coat of paint is applied after the job is complete to seal the crevice between the components and the lead or nut of the bolt, and as an extra protection of surfaces exposed (perhaps accidentally) during the fixing process.

Danger of contact corrosion, and other points, must also be considered in selecting materials or protective coatings for junction units. On functional and aesthetic grounds, for example, unprotected steel screws should never be used to join parts which are already well-protected' against corrosion. This is particularly true if the part must be dismantled periodically (e.g. for repair). Galvanized, aluminized or enamelled steels or anodized aluminium lose their attractive surface appearance in a short time if 'rusty' water flows across them and leaves traces behind. If the surface protection of nuts or bolts (especially small ones) poses problems and there is no sufficiently long-lived coating available, it is worthwhile to replace them in a more noble or corrosion-passive alloy, as long as it does not endanger the components by contact corrosion (and the danger of this is usually low, as was discussed in Chapter 7.3.2). Galvanized steel normally requires galvanized bolts (though chromed ones also provide good corrosion resistance), aluminium-coated steel can be fixed using galvanized, cadmium-coated, or chromed bolts, and these same junction units can also be used for aluminium if there are none available in suitable aluminium alloys, etc. It should not be necessary to mention that the protection by the coating on the junction unit should be of the highest possible quailty (i.e. sufficiently thick coating, passivation, supplementary paint layer).

7.3.4 Process technology considerations

In addition to the points raised so far, some consideration must also be given to the possible processes for application of coatings of the desired quality, and to the problems of which can arise if no provision is made for subsequent measures to repair the coating.

Two groups of factors are involved in considering the techniques available: the technical equipment of the manufacturer and the cost of the protection, and the structural configuration of the product to be protected. These questions are very important, in that they affect not only the end quality of the product but also the cost of its protection.

It is scarcely satisfactory if, once the corrosion engineer has decided on a certain protective coating, the necessary equipment is not available in the factory to produce the coating, and its purchase is not economical and, further, contracting-out of the coating production is unprofitable because of e.g. high transport costs. As an example: the corrosion expert specifies the optimum corrosion protection of a steel scaffolding as hot galvanizing followed by painting. The manufacturer cannot perform hot galvanizing in his plant, however, and the nearest commercial galvanizing plant is some distance away. A simple calculation will show whether it will pay to make allowance for contracting-out of the galvanizing despite the extra costs it will involve, or whether it will be better to settle for another solution, such as metal spraying by hand of a zinc or aluminium layer, and subsequent painting, with acceptance that maintenance may now be needed sooner than initially planned. It is also senseless to demand coating with a stoving lacquer if the plant is not equipped to do this, especially since there are virtually equivalent systems available nowadays which dry rapidly at environmental temperatures.

The labour content involved in surface protection should also be considered. Very complicated operations are not only expensive, but their quality is also difficult to standardize or control.

The size and shape of the components should also be taken into account in designing corrosion protection measures. There are frequent cases where neglect of this has meant that the planned protection can be used with difficulty or not at all. A piece of aluminium cannot be anodized if it is bigger than the available anodizing tank; the protective system must then be applied by performing the operation twice (i.e. once each end), which makes the operation more expensive and leaves an unattractive mark in the middle. It is generally impossible to immerse a hollow component, of which only the outside is to be galvanized, into a hot galvanizing bath, because of the hydrostatic pressures involved. Small parts (e.g. gear wheels, or axles for precision equipment) which are required to be highly resistant to corrosion for functional reasons, and which are exposed to very aggressive atmospheres, such as parts of regulatory equipment, cannot usually be protected by galvanized coatings because of the process technology involved. They are not properly protected because their complicated shapes usually result in insufficient layer thicknesses on parts of the component, and should instead be produced from passive metals which require no protective coating.

It is generally well-known that sharp edges, places which are difficult to reach, and complicated shapes pose problems to practically every protection method, and lead to non-uniform protection quality. The simpler the shape and the more straightforward and accessible the surface, the cheaper and better is the coating which can be applied to it. (Special notes on desirable structural considerations to facilitate corrosion protection may be found in e.g. van Oeteren [2]).

In considering protection of long-lived products, attention should be paid to the likely necessity of repair and the costs associated with this. This is often overlooked, especially on buildings, and usually makes their maintenance more difficult than it might have been, since special scaffolding must then be erected to reach all the deteriorated surfaces. Suspended platforms, for example, can be used if they are planned for from the outset, and the necessary runways, anchoring points, etc., are built in, and this saves money in the long term. Other practical examples may be found in [2]. It must be assumed that the corrosion engineer will have ensured the accessibility of all surfaces in danger of corrosion when the structure was being planned.

7.3.5 Performance of corrosion protection methods during manufacture, and assessment of the quality of protection

A protective system can only produce good results i.e. the expected quality of protection, if all operations which affect the end result have been carefully performed. Surface examination of completed protective coatings can lead to erroneous conclusions, since a good appearance, freedom from pores, and a check that the prescribed minimum thickness is present provide no guarantee of the required life. It is not uncommon to find that an apparently faultless paint layer has under-rusted and peeled off after a short exposure to the atmosphere.

It is usually not difficult to detect the reason for such failures. It is poor pre-treatment, which allows soluble species to be trapped under the paint layer, so that the underlying metal can corrode easily after penetration of water and disruption of the adhesion, which is already weak in such cases.

Similarly, other unexpected corrosion phenomena can be found. Blister formation may be connected with e.g. insufficient hardness of the intermediate layer; galvanized coatings can crack due to unexpected internal stresses, etc. These are all found after the damage has been done, however, and the damage cannot always be made good again, or may require very expensive additional work. Measures can only be taken to try to prevent a recurrence of the problem.

Suitable controls, which should be as simple as possible, on production processes must be instituted to prevent such unexpected occurrences. It invariably pays dividends to inspect the results of all important operations during the surface protection process. It is usually desirable to examine the quality of surface pre-treatment, and the individual layers in a multilayer protective system. The final quality of the protective coating should be assessed in the light of these intermediate tests as well as the (usually not especially meaningful) values of thickness, appearance, hardness, etc.

The multiplicity of protective methods makes it virtually impossible to enumerate all the desirable measurements, but mention should be made of a few measurements which are frequently overlooked.

Product specifications contain exact definitions of the surface state after sand-blasting as a preparation for painting or thermal spraying. In many cases, however, the specifications fail to require that the performance of the subsequent operation must follow soon enough for no rusting to have occurred; yet this rusting is usually very dangerous. Paints dry at different rates, especially if they are applied in the open air, depending on weather conditions. If the next layer is applied too soon, irreparable damage is done. Painting is not always supervised by an experienced and reliable foreman who can decide the correct time for application of the next layer without some objective test of the degree of drying achieved. A simple control method might be specified here, to help the less experienced; this might be no more complicated than a measurement of the hardness of the paint layer using a pencil.

It is incorrect to assume that mechanized or automatic methods of surface protection production always guarantee a high and uniform coating quality. Much depends on the operator of the machinery. Anger (if not more serious feelings) is aroused if it is found subsequently that an expensive multi-layer coating has pores in it, or peels off for some (usually trivial) reason; e.g. the degreasing bath has worked poorly because of insufficient supervision of its chemical composition, or one of the electrolytes is polluted by foreign cations (e.g. Fe^{2+} in a copper bath, or Cu^{2+} in a nickel bath), or because some of the products have been left 'forgotten', lying in the bath with no current running, etc. It is the task of the corrosion engineer to take preventitive measures against such as these irregularities in process control might produce, perhaps by specifying particular inspections after some steps.

Not only those protective methods which rely on the use of different coatings and their combination should be inspected during production; the same applies to all protection methods. Passive metals, such as stainless steels, should be systematically tested (since confusion between austenitic Cr–Ni-steels and the less resistant ferritic chromium steel is not uncommon). The reliability of air-conditioning equipment which ensures that an indoor atmosphere is favourable for protection should be checked frequently by exact measurements of temperature and humidity at different points in the room.

7.4 Protection measures for products in the period before use

A product which has been protected against the corrosive action of the atmosphere at its final destination does not always survive the interval between its completion and its commissioning in the same state. During storage and transport, quite different conditions may arise against which protection was never contemplated.

An example of this: An underground pipeline was required to carry fuel to aircraft refuelling points at an airport. Since the fuel must be kept extremely pure, the pipe was produced as a series of sealed cylinders with shiny internal surfaces, protected by a vapour phase inhibitor. During assembly and burial, it was unprotected, though it was intended to come into use some months later. It was then discovered that the pipeline had rusted badly on its internal surfaces while underground, and expensive cleaning was needed before it could be used without endangering the safety of the aircraft. The corrosion engineer succeeded in restoring the pipeline here, and specified renewal of temporary protection after each operation (e.g. by filling the restored part of the pipeline with an aqueous solution of sodium nitrite).

Similar cases of corrosion are found during construction of large-scale industrial plants. There should be temporary protection applied to surfaces of e.g. unalloyed construction steel, which is sufficiently resistant when installed but must be installed with the least possible corrosion products on its surface.

Again, similar considerations apply in the design of temporary protection which is to be used on a product while it is in the factory. Parameters involved in decisions here include the probable longest duration of the necessary protection, the likely environmental conditions around the product during this time, and possible effects on the temporary protection of handling during movement of the product, during handling, and from time of assembly to commissioning.

The duration of temporary protection of the product may be quite different for different individual products from the same line. The protection should be made very long-lasting if transport by sea or construction of large plants are involved. There are clearly different requirements made on the protection of an automobile which is to be sold in the country of manufacture and of one which may travel by rail to a depot after delivery by sea, and may then stand for some months before sale.

Even after very careful transport, primitive storage conditions in tropical countries may endanger expensive equipment, and so storage periods of an extra month or more should be allowed for when planning temporary protection. From experience, loading and temporary storage methods in coastal towns are especially hazardous. It frequently occurs that a product will stand for a week in the open air on the dockside, and thus in a very aggressive marine atmosphere, before it is loaded.

The choice of a protective system for this period of the life of technical apparatus is not simple, and must not be neglected. It is always better to use more powerful temporary protection than to be faced with repair (which will probably be expensive) or replacement of parts damaged because the danger of corrosion was under-estimated.

Temporary protection usually involves two factors; a temporary preservative and suitable packing methods. From the point of view of corrosion, the packing is mainly to maintain the temporary preservative action, and so both steps are equally important. The temporary protection should cover the period between product completion and its introduction to service, and the packing should ensure that this can be done with an undiminished protective action over this period. If this cannot be achieved, renewal of the temporary protection measures should be arranged; this might have to be done for economic reasons or because the protection is required to last for relatively long periods. This renewal, which might be performed e.g. in the country of destination, should be such that a predetermined state is attained, or a predetermined amount of preservative is added. Even measures such as this are cheaper than allowing atmospheric corrosion to occur, with its concomitant danger of impairment of product functions.

7.5 Ideas on choice of measures for atmospheric corrosion protection

7.5.1 General classification of atmospheric aggressiveness

Since the corrosive nature of an atmosphere is determined mainly by local conditions, with a secondary, though not negligible, role played by the geographical macro-climate, it seems suitable to use as a basis for atmospheric classification the two most important factors for corrosion promotion i.e. the time of surface wetting and the intensity of stimulator action (see also Chapter 7.2). In this way, a five-step system can be devised, on which can be based specifications of surface protection quality (Table 14).

Table 14. Classification of aggressive natures of different atmospheres

Degree of aggressiveness:	1	2	3	4	5
		Corrosion removal (μm year^{-1})			
Unalloyed carbon steel	<0·01	0·01 to 0·1	0·1 to 25	25 to 50	>50
Copper	0	0 to 0·1	0·1 to 3	3 to 5	>5
Aluminium	0	0	0 to 0·1	0·1 to 1	>1
Zinc	<0·01	0·01 to 0·1	0·1 to 5	5 to 10	>10

Table 15. Usefulness of metals (as pore-free coatings) if corrosion products are undesirable. ($+$ = acceptable, $-$ = unacceptable). The columns 1 to 5 correspond to the aggressiveness of the atmosphere according to Table 14.

	1	2	3	4	5
Carbon steel, cast iron, low-alloy steels	−	−	−	−	−
Stainless Cr–, Cr–Mn–, and Cr–Mn–Ni–steels	+	+	+	+	−
Stainless Cr–Ni(or Mo)–steels	+	+	+	+	+
Copper and copper alloys	+	−	−	−	−
Aluminium (also as a coating)	+	+	+[a]	−	−
Zinc, zinc alloys and zinc coatings	+	+[b]	−	−	−
Cadmium coatings	+	+[b]	−	−	−
Lead, tin (also as coatings)	+	−	−	−	−
Silver (also as a coating)	+[c]	+[c]	+[c]	−	−
Gold, platinum, rhodium, titanium, zirconium, tantalum, niobium (also as coatings)	+	+	+	+	+
Nickel and niclel coatings	+[c]	+[c]	+[c]	−	−
(Copper–)nickel–chromium coatings	+	+	+	+	+
Chromium coatings	+	+	+	+	+

Notes: a: in absence of chloride.
　　　b: when chromated.
　　　c: on absence of sulphur compounds.

The five steps in the scale used here are defined by the expected corrosion loss of many widely-used metals, so that the danger of corrosion can also be estimated from degree of corrosion product formation (Table 15). Using this scale of aggressiveness, the scheme in Table 16 can be used to allow for general macro- and micro-climatic factors, and variations in exposure.

Table 16. Corrosive action (using the scale from Table 14) of atmospheres in different climatic zones, with respect to specific local conditions

Type of impurity*	Exposure†	Macroclimatic zone			
		dry tropical	cold	temperate	damp tropical
$SO_2 < 0.01\,g\,m^{-2}\,d^{-1}$	a	2	2	3	4
$NaCl < 0.3\,g\,m^{-2}\,d^{-1}$	b	2	2	3	4
	c	1	1	2	3
$SO_2\,0.01$ to $0.1\,g\,m^{-1}\,d^{-1}$	a	3	3	4	5
$NaCl < 0.3\,g\,m^{-2}\,d^{-1}$	b	3	3	4	5
	c	2	2	3	4
$SO_2 > 0.1\,g\,m^{-2}\,d^{-1}$	a	4	4	5	5
$NaCl\,0.3$ to $2\,g\,m^{-2}\,d^{-1}$	b	4	4	5	5
	c	3	3	4	5
$NaCl > 2\,g\,m^{-2}\,d^{-1}$	a	2	3	4	5
(i.e. marine atmosphere)	b		3	4	5
	c		2	3	4

Specific impurities must be examined individually.
* See Chapter 6.
† a: Free weathering; b: protected only from precipitation and direct sunlight; c: uncontrolled indoor atmosphere (e.g. unheated warehouse).
Indoor atmospheres with controlled temperature and humidity must be examined individually, but are usually Class 1.

7.5.2 Recommended minimum thicknesses for metallic coatings

These are summarized in Table 17, in terms of the aggressiveness of the atmosphere as classified in Tables 14 to 16. The layer thicknesses quoted are for minimum lives of 5 to 10 years.

7.5.3 Protective coatings for particularly aggressive conditions

Extremely corrosive conditions can occur around chemical plants and in similar areas under certain conditions, and not simply the stimulator species usually present are involved in the corrosion in these atmospheres. The surface must be protected against temporary presence of acid, alkali, etc. A review of possible protection systems for steel in these conditions is given in Table 19. Thermally-sprayed aluminium layers (0.2 to 0.3 mm thick) or zinc silicate

Table 17. Recommended coating thicknesses (µm) for use in different aggressive atmospheres and conditions

Material	Coating	Process	1	2	3	4	5
Steel, cast iron	Zinc	Electrolytic	7	7	15	30	
		Immersion		20		40 to 100	
		Thermal spray			60	100	200
	Cadmium	Electrolytic	7	7	15	30	
	Tin	Electrolytic ⎱ Immersion ⎰	10	10	20	30	30
	Aluminium	Thermal spray			100	200	300
	Lead	Electrolytic			20 to 30		
		Immersion			15 to 30		
	Chromium	Diffusion		30		30	30
		Electrolytic (hard or matt layers)		30		50	70
							100
	Nickel	Electrolytic	10	20	30	40	
		Chemical			see Table 18		
	Copper–nickel(–chromium)	Electrolytic					
	Copper–silver ⎱ Nickel–silver ⎰	Electrolytic (silvering of an intermediate layer)	10–5		15–10 to 25		
	Phosphating + oil or grease (or wax)	Chemical	100 to 300 mg dm^{-2}				not suitable

Material group	Coating	Method	1	2	3	4	5
Cement		Painting			100	200	400
Inorganic zinc silicate paint		Painting			150		
Enamel					not specified		
Copper, copper alloys	Nickel	Electrolytic		5			
		Chemical		10	20	30	
	Nickel–chromium	Electrolytic			see Table 18		
	Tin	Electrolytic / Immersion	3	3	7	15	15
	Silver	Electrolytic	5	5	10	10	15
	Gold	Electrolytic		1 to 20		10 to 20	15
Aluminium, and its alloys	Chromium	Electrolytic (hard or matt)		30	(depending on the resistance to abrasion required)	50	100
	Copper–nickel(–chromium)	Electrolytic	3/10/0.25	5/10/0.25	10/20/0.25	10/20/0.25	10/30/0.25
	Oxide	Electrolytic	5	10	15	25	25 (not suitable for extreme conditions)
Zinc alloys	See Table 18 for electrolytic (copper–)nickel–chromium layers						
	Copper–nickel	Electrolytic	8/8	8/15	8/25	not suitable	
	Phosphating	Chemical			not specified		
	Chromating	Chemical			not specified		

1 to 5 refers to the atmospheric aggressiveness, defined in Table 14.

Table 18. Recommended layer thicknesses (μm) of decorative electrolytic bright chrome coatings (following the ISO recommendations, and using the atmosphere classification from Table 14)

Substrate	Nickel type	Chromium type and thickness (μm)	Atmosphere class 1	2	3	4(5)
Steel	Bright nickel	normal 0·3	Ni 10 Cu 10 Ni 5	Ni 20 Cu 20 Ni 10	Ni 40 Cu 20 Ni 30	not desirable
		crack-free 0·8	Ni 10 Cu 10 Ni 5	Ni 20 Cu 10 Ni 10		
		micro-cracked 0·3	Ni 10 Cu 10 Ni 5	Ni 20 Cu 20 Ni 10	Ni 30 Cu 20 Ni 20	Ni 40 Cu 20 Ni 25
	Matt or semi-polish (<0·005% S) mechanically polished	normal 0·3	Ni 10 Cu 10 Ni 5	Ni 20 Cu 20 Ni 10	Ni 30 Cu 15 Ni 25	Ni 40 Cu 20 Ni 30
		crack-free	Ni 10 Cu 10 Ni 5	Ni 20 Cu 20 Ni 10		
		micro-cracked 0·3	Ni 10 Cu 10 Ni 5	Ni 20 Cu 20 Ni 10	Ni 25 Cu 15 Ni 20	Ni 30 Cu 20 Ni 25
	Double or triple nickel	normal 0·3	Ni 10 Cu 10 Ni 5	Ni 20 Cu 20 Ni 10	Ni 30 Cu 15 Ni 25	Ni 40 Cu 20 Ni 30
		crack-free 0·8	Ni 10 Cu 10 Ni 5	Ni 20 Cu 20 Ni 10	Ni 30 Cu 15 Ni 25	Ni 40 Cu 20 Ni 30
		micro-cracked (microporous) 0·3*	Ni 10 Cu 10 Ni 5	Ni 20 Cu 20 Ni 10	Ni 25 Cu 15 Ni 20	Ni 30 Cu 20 Ni 25

Substrate	Nickel type	Condition					
Zinc alloys	Bright nickel	normal 0·3	Cu 8 Ni 8	Cu 8 Ni 15	Cu 8 Ni 25	Cu 8 Ni 35	not desirable
		crack-free	Cu 8 Ni 8	Cu 8 Ni 15	Cu 8 Ni 25	Cu 8 Ni 35	Cu 8 Ni 35
		micro-cracked 0·3	Cu 8 Ni 8	Cu 8 Ni 15	Cu 8 Ni 25	Cu 8 Ni 25	Cu 8 Ni 25
	Matt or semi-matt (<0·005% S) polished	normal 0·3	Cu 8 Ni 8	Cu 8 Ni 15	Cu 8 Ni 25	Cu 8 Ni 25	Cu 8 Ni 35
		crack-free 0·8	Cu 8 Ni 8	Cu 8 Ni 15	Cu 8 Ni 25	Cu 8 Ni 25	Cu 8 Ni 35
		micro-cracked 0·3	Cu 8 Ni 8	Cu 8 Ni 15	Cu 8 Ni 20	Cu 8 Ni 20	Cu 8 Ni 25
	Double or triple nickel	normal 0·3	Cu 8 Ni 8	Cu 8 Ni 15	Cu 8 Ni 25	Cu 8 Ni 25	Cu 8 Ni 35
		crack-free 0·8	Cu 8 Ni 8	Cu 8 Ni 15	Cu 8 Ni 25	Cu 8 Ni 25	Cu 8 Ni 35
		micro-cracked	Cu 8 Ni 8	Cu 8 Ni 15	Cu 8 Ni 20	Cu 8 Ni 20	Cu 8 Ni 25
		(microporous) 0·3*					
Copper and copper alloys	Bright nickel	normal 0·3	Ni 5	Ni 10	Ni 20	Ni 30	Ni 40
		crack-free 0·8	Ni 5	Ni 10	Ni 20	Ni 30	
		micro-cracked 0·3	Ni 5	Ni 10	Ni 20	Ni 20	Ni 25
	Matt or semi-matt (<0·005% S) polished	normal 0·3	Ni 5	Ni 10	Ni 20	Ni 20	Ni 30
		crack-free 0·8	Ni 5	Ni 10	Ni 20	Ni 20	
		micro-cracked 0·3	Ni 5	Ni 10	Ni 20	Ni 20	Ni 25
	Double or triple nickel	normal 0·3	Ni 5	Ni 10	Ni 20	Ni 30	Ni 30
		crack-free 0·8	Ni 5	Ni 10	Ni 20	Ni 30	Ni 30
		micro-cracked	Ni 5	Ni 10	Ni 20	Ni 20	Ni 25
		(microporous) 0·3*					

* With a nickel sealing layer below the chromium.

Table 19. Resistant protective coatings for steel in extremely aggressive atmospheric conditions

Aggressive action by:	Recommended protection systems:
SO_2, dust, water, high humidity	Thermally sprayed zinc layer (0·1 mm) + 2 or 3 coats of chemically resistant paint (e.g. based on chloroprene) Thermally sprayed aluminium layer (0·2 mm) + 2 or 3 coats of a chemically resistant paint. Hot galvanizing (0·05 mm) + chemically resistant paint Zinc silicate paint (0·1 to 1 mm) + organic paint
NaCl, water, high humidity	Cement paint (0·5 to 1 mm) + chemically resistant organic paint (e.g. chloroprene) Zinc silicate paint (0·1 to 1 mm) + organic paint
NH_3, high humidity	Cement paint (0·1 to 1 mm) + alkali-resistant organic paint (e.g. epoxide, chloroprene, etc.)
H_2SO_4, wetting	Covering with lead Zinc silicate paint with an organic surface coating (chloroprene, poly-vinyl chloride, etc.)
Prolonged wetting	Cement paint (0·5 to 1 mm) with an organic surface coating (e.g. vinyl copolymer, chloroprene)

paints are the most suitable methods for temporary protection against higher temperatures and harsh atmospheric conditions.

7.5.4 Suggested temporary preservative-packing systems

The reliability of different combinations of methods of temporary protection and packing for different types of atmospheres and time requirements is summarized in Table 20.

It must be noted that Table 20 is merely a guide. It is assumed that tested materials of undoubted quality are used, and that process technology faults, such as incomplete cleaning of the surface, can be ignored.

Table 20. Review of the reliability of different combinations of temporary protection media with different packaging. (Packaging classification as in Table 21, Atmosphere classification as in Table 14)

Atmosphere type	Packing type	Protection period	Protection medium						
			inhibiting oil	inhibiting grease	inhibiting wax	stripping lacquer (thin)	stripping material (thick)	inhibiting paper	none
1–2	I	up to 6 months	×	×	+	○	+	×	○
	II		+	+	+	○	+	+	○
	IIIa		+	+	+	+	+	+	+
	IIIb		+	+	+	+	+	+	+
	IIIc		+	+	+	○	+	+	○
3–4	I		○	○	×	○	+	×	○
	II		+	+	+	○	+	+	○
	IIIa		+	+	+	+	+	+	+
	IIIb		+	+	+	+	+	+	+
	IIIc		+	+	+	○	+	+	○
1–2	I	up to 2 years	(+, ○)	×	×	○	+	×	○
	II		+	+	+	○	+	+	○
	IIIa		+	+	+	+	+	+	+
	IIIb		+	+	+	+	+	+	+
	IIIc		+	+	+	○	+	+	○
3–4	I		○	○	×	○	+	?	○
	II		+	+	+	○	+	+	○
	IIIa		+	+	+	+	+	+	+
	IIIb		+	+	+	+	+	+	+
	IIIc		+	+	+	○	+	+	○
1–2	I	more than 2 years	○	○	×	○	+	×	○
	II		+	+	+	○	+	+	○
	IIIa		+	+	+	+	+	+	+
	IIIb		+	+	+	+	+	+	+
	IIIc		○	+	+	○	+	+	○
3–4	I		○	○	○	○	○	○	○
	II		○	+	+	○	+	+	○
	IIIa		+	+	+	+	+	+	+
	IIIb		+	+	+	+	+	+	+
	IIIc		○	+	+	○	+	+	○

+ : Applicable; × : applicable if used with a watertight outer packing; ○ : not applicable.

Table 21. Description of packaging types (see Table 20)

Designation	Description
I unsealed	Protected against rain, mist, sea-water which has been atomized and formed an aerosol, and mechanical pollutants (usually dust)
II water-tight seal	Protected against infiltration of liquid water
III gas-tight seal	Protected against the infiltration of liquid water, dust, water vapour, and all gaseous components of the atmosphere a. using a sufficient (calculated) amount of dessicant (e.g. silica gel) b. using sufficient dessicant to render harmless any enclosed water, and low water permeability of the packing material c. without dessicant, and impermeable to water; dried out, for example by evacuation and refilling with dry air.

References

Chapter 1

1. H. H. Uhlig, *Corrosion* **6**, 29 (1950).
2. K. Daeves, K. Trapp, *Stahl und Eisen* **59**, 169 (1937).
3. K. Hauffe, *Oxidation of Metals*, Plenum, New York, 1965.
4. U. R. Evans, *The Corrosion and Oxidation of Metals*, Arnold, London, 1960.
5. N. D. Tomashov, *Theory of Corrosion and Protection of Metals*, Macmillan, New York, 1966.
6. H. Kaesche, *Die Korrosion der Metalle (Corrosion of Metals)*, Springer Verlag, Berlin, 1966.
7. U. F. Franck, *Werkstoffe und Korrosion* **14**, 367 (1963).

Chapter 2

1. J. Hann, *Lehrbuch der Meteorologie (Textbook of Meteorology)*, Keller, Leipzig, 1939.
2. V. Conrad, *Handbuch der Klimatologie I*, Gebr. Borntrager, Berlin, 1936: *Die Klimatologischen Elemente und ihre Abhängigkeit von terrestrischen Einflussen (Climatological Factors and Their Dependence on Terrestrial Influences)*.
3. H. Kaesche, *Werkstoffe und Korrosion*, **15**, 379 (1964).
4. A. Kutzelnigg, *Werkst. Korros.*, **8**, 492 (1957).
5. F. M. Clark, *Electr. Engng.*, 671 (1937).
6. D. Čermáková-Knotková, *Dissertation*, Staatl. Forschungsinst. für Materialschutz, Prague, 1967.
7. T. K. Ross, B. G. Callaghan, *Nature*, **211**, 25 (1966).
8. G. Oelsner, *Werkst. Korros.*, **15**, 305 (1964).
9. J. L. Rosenfeld, *Atmosfernaja Korrozija metallov (The Atmospheric Corrosion of Metals)*, Izd. Akad. Nauk SSSR, Moscow, 1966.
10. K. Bartoň, *Werkst. Korros.*, **9**, 547 (1958).
11. E. Beránek, K. Bartoň, K. Smrček, I. Sekerka, *Coll. Czech. Chem. Comm.*, **22**, 368 (1957).
12. R. Geiger, *Das Klima der bodennahen Luftschicht (The Climate of Air Layers near the Earth)*, Vieweg, Braunschweig, 1950.
13. K. A. van Oeteren, *Konstruktion und Korrosionschutz (Construction and Corrosion Protection)*, Curt R. Vincentz Verlag, Hanover, 1967.
14. J. L. Rosenfeld: *Korrosija i zaschtschita metallov (Corrosion and Protection of Metals)*, Metallurgia, Moscow, 1969, Chapter 4: 'Corrosion and construction concepts'.
15. R. Schropp, *Gesundheits-Ingenieur*, 729 (1931).

Chapter 3

1. D. M. Brasher, *British Corros. J.*, **2**, 95 (1967); **4**, 122 (1969).
2. e. g. K. Hauffe, *Oxidation of Metals*, Plenum, New York, 1965.
3. R. Bartoníček and co-workers, *Koroze a protikorozní ochrana kovů (Corrosion and Protection of Metals)*, Academia Praha, 1966, pp. 148 to 162.

184

4. U. R. Evans, *Nature,* **164**, 909 (1949).
5. U. R. Evans, *The Corrosion and Oxidation of Metals,* Arnold, London, 1960, pp. 844–845.
6. N. Cabrera, N. Mott, *Rept. Progr. Phys.,* **12**, 163 (1949).
7. T. N. Rhodin, *J. Am. Chem. Soc.,* **72**, 5102 (1950).
8. R. K. Hart, *Proc. Royal Soc. (London),* **A236**, 68, 81 (1956).
9. P. T. Landsberg, *J. Chem. Phys.,* **23**, 1079 (1955).
10. C. Weissmantel, *Werkst. Korros.,* **13**, 682 (1962).
11. U. F. Franck, *Werkst. Korros.,* **14**, 367 (1963).
12. M. Pourbaix, *Thermodynamics of Dilute Aqueous Solutions,* Arnold, London, 1949.
13. P. J. Sereda, *ASTM Bulletin,* **246**, May 1960.
14. A. M. Zinewich, E. N. Sergeeva, J. N. Mikhailovski, W. B. Serafimovich, *Zashch. Metall.,* **6**, 333 (1970).
15. H. Kaesche, *Werkst. Korros.,* **15**, 379 (1964).
16. K. Bartoň, *Werkst. Korros.,* **9**, 547 (1958).
17. A. Buckowiecki, *Schweizer Archiv Angew. Wiss. Techn.,* **23**, 97 (1957).
18. H. Schwarz, *Werkst. Korros.,* **16**, 93, 208 (1965).
19. K. Bartoň, unpublished data.
20. G. K. Berukschtis, G. B. Klark, *Korrosija i zaschtschtita Splavov (Corrosion and Protection of Alloys),* Metallurgia, Moscow, 1965.
21. J. L. Rosenfeld: '*Atmosfernaja korrosija metallov*' ('*Atmospheric Corrosion of Metals*'), Izd. Akad. Nauk SSSR, Moscow, 1960.
22. J. L. Rosenfeld, K. A. Zhigalova, *Uskorenyje metody korrosionnych ispitanii metallov (Accelerated Corrosion Testing Methods),* Metallurgia, Moscow, 1966.
23. J. L. Roich, *Dokl. Akad. Nauk SSSR,* **108**, 372 (1956); **111**, 362 (1956).
24. W. W. Gerasimov, J. L. Rosenfeld, *Isv. Akad. Nauk SSSR,* **7**, 279 (1956).
25. R. Ch. Burstein, A. N. Frumkin, *Dokl. Akad. Nauk SSSR,* **32**, 327 (1941).
26. J. L. Rosenfeld, K. A. Zhigalova, *Dokl. Akad. Nauk SSSR,* **99**, 137 (1954).
27. K. Bartoň, Ž. Bartoňová, *Werkst. Korros.,* **21**, 25 (1970).
28. H. Schwarz, 57. *Veranstaltung EFK,* Prague, 1971, pp. 13 to 56: 'Schutz von Stahlkonstruktion gegen atmospherische Korrosion' ('Protection of steel structures against atmospheric corrosion').
29. E. Beránek, K. Bartoň, K. Smrček, I. Sekerka, *Coll. Czech. Chem. Comm.,* **22**, 368 (1957).
30. K. Schwabe, *Werkst. Korros.,* **15**, 70 (1964).
31. K. E. Heusler, *Z. Elektrochem.,* **62**, 582 (1958).
32. J. O'M. Bockris, *Electrochim. Acta,* **7**, 293 (1962).
33. W. J. Lorenz, *Corrosion Science,* **5**, 121 (1965).
34. G. M. Florianovich, L. A. Sokolova, Ja. M. Kolotyrkin, *Electrochim. Acta,* **12**, 879 (1967).
35. *Proc. 3rd Int. Congr. Metallic Corrosion,* Moscow, 1966, Volume 1.
36. W. Feitknecht, *Chimia,* **6**, 3 (1952).
37. J. F. Henriksen, *Corros. Sci.,* **5**, 573 (1969).
38. G. Schikorr, *Werkst. Korros.,* **15**, 457 (1964).
39. J. L. Rosenfeld: '*Atmosfernaja korrosija metallov*' ('*Atmospheric Corrosion of Metals*'), Izd. Akad. Nauk SSSR, Moscow, 1960.
40. G. G. Koschelev, G. B. Klark, *Trudy Inst. Fiz. Chim. AN SSSR,* **8**, 84 (1960).
41. K. Bartoň, Ž. Bartoňová, *Werkst. Korros.,* **20**, 216 (1969).
42. K. Bartoň, E. Beránek, *Werkst. Korros.,* **10**, 377 (1959).
43. W. McLeod, R. R. Rogers, *Corrosion,* **22**, 143 (1966).
44. D. Čermáková-Knotková, J. Vlčková, *Werkst. Korros.,* **21**, 16 (1970).
45. T. K. Ross, B. G. Callaghan, *Electrochim. metall.,* **2**, 22 (1967).
46. E. Herzog, *Métaux,* **500**, 133 (1967).
47. G. Schikorr, *Werkst. Korros.,* **14**, 69 (1963).
48. K. Bartoň, D. Kuchyňka, *Corros. Sci.,* **11**, 937 (1971).
49. K. Bartoň, Ž. Bartoňová, *Werkst. Korros.,* **21**, 25 (1970).
50. K. Bartoň, D. Cermáková, *Werkst. Korros.,* **15**, 374 (1964).
51. T. Biestek, J. Niemec, *Prace Inst. Mechaniki Prec.,* **14**, 58 (1968).
52. P. M. Aziz, H. P. Godard, *Corrosion,* **15**, 529t (1959).
53. K. Bartoň, Ž. Bartoňová, *Proc. 3rd. Int. Congr. Metallic Corrosion, Moscow,* 1966, Vol. 4, p. 483 (English edition).
54. R. Grauer, *Werkst. Korros.,* **20**, 991 (1969).
55. G. Oelsner, *Werkst. Korros.,* **15**, 305 (1964).
56. K. Bartoň, *Werkst. Korros.,* **9**, 547 (1958).

57. W. H. J. Vernon, *Trans. Faraday Soc.*, **23**, 162 (1927); **27**, 264 (1931); **29**, 35 (1933).
58. E. Längle, *Schweizer Archiv. Angew. Wiss. Tech.*, **34**, 109, 147 (1968).
59. K. Bartoň, Ž. Bartoňová, *Werkst. Korros.*, **20**, 87 (1969).
60. E. K. Osche, J. L. Rosenfeld, *Elektrokhimiya*, **2**, 1200 (1968).
61. D. Čermáková-Knotková, *Dissertation*, Staatl. Inst. für Materialschutz, Prague, 1967.
62. T. K. Ross, B. G. Callaghan, *Corros. Sci.*, **6**, 337 (1966).
63. J. Honzák: 57. Veranstaltung EFK, Prague, 1971: 'Schutz von Stahlkonstruktion gegen atmospharische Korrosion ('Protection of steel structures against atmospheric corrosion').
64. H. Kaesche, *Die Korrosion der Metalle (Corrosion of Metals)*, Springer Verlag, Berlin, 1966, Chapter 7.
65. R. H. Copson, *Proc. ASTM*, **45**, 554 (1945); **52**, 1005 (1952).
66. G. H. Becker, D. Dhingra, C. Thoma, *Arch. Eisenhütten.*, **40**, 341 (1969).
67. *6th Report of the Corrosion Committee*, Iron and Steel Institute, London, 1959.
68. J. F. Stanners, *British Corros. J.*, **5**, 117 (1970).
69. S. Odén, *Korrosionsnämnd Bull. (Stockholm)*, **45**, (1965).
70. T. K. Ross, B. G. Callaghan, *Nature*, **211**, 25 (1966).
71. W. Schwenk, H. Ternes, *Stahl und Eisen*, **88**, 318 (1968).
72. A. I. Golubjev, M. Kadyrov; *Proc. 3rd Int. Congr. Metallic Corrosion, Moscow*, 1966, Vol. 4, p. 522 (English Edition).
73. H. Guttman, P. J. Sereda, *ASTM Special Tech. Publ.*, **435**, 336 (1968).
74. G. K. Berukschtis, G. B. Klark: *Korrosija i zaschtchita Splavov (Corrosion and Protection of Alloys)*, Metallurgia, Moscow, 1965, p. 232.
75. P. J. Sereda, *Materials Research and Standards*, 719 (1961).
76. N. D. Tomashov, A. A. Lokotilova, *Zav. lab.*, **24**, 4, 425 (1958).
77. P. J. Sereda, *Ind. Engng. Chem.*, **52**, 157 (1960).
78. N. S. Patterson, L. Hebbs, *Trans. Faraday Soc.*, **27**, 277 (1931).
79. R. S. J. Preston, B. Sanyal, *J. Appl. Chem.*, **6**, 26 (1956).
80. K. Bartoň, Ž. Bartoňová, *Werkst. Korros.*, **21**, 85 (1970).
81. J. C. Hudson, *Werkst. Korros.*, **13**, 363 (1964).
82. K. Bartoň, *Z. phys. Chemie*, **71**, 72 (1965).
83. H. Guttman, *ASTM Special Tech. Publ.*, **435**, 223 (1968).
84. V. Marek, unpublished data.
85. J. C. Hudson, J. F. Stanners, *J. Appl. Chem.*, **9**, 673 (1953).
86. T. Sydberger, N. G. Vannerberg, *Corros. Sci.*, **12**, 775 (1972).
87. N. G. Vannerberg, T. Sydberger, *Corros. Sci.*, **10**, 43 (1970).
88. B. Heimler, N. G. Vannerberg, *Corros. Sci.*, **12**, 579 (1972).
89. J. R. Duncan, *Werkst. Korros.*, **25**, 420 (1974).
90. J. R. Duncan, D. J. Spedding, *Corros. Sci.*, **13**, 69, 881 (1973); **14**, 241 (1974).
91. P. V. Strekalov, Y. N. Mikhailovski, *Zaschtsch. Metall.*, **8**, 573 (1972).
92. G. Schikorr, *Aluminium (BRD)*, **43**, 108 (1967).
93. H. Okada, H. Shimada, *Corrosion (Houston)*, **30**, 97 (1974).
94. J. F. Stanners, *Proc. 4th Int. Congr. Metallic Corrosion, NACE, Houston*, 1972, p. 419.
95. F. H. Haynie, J. B. Upham, *Materials Protection and Performance*, **9**(8), 35 (1970); **10**(11), 18 (1971).
96. D. Knotková, B. Bošek, 'Corrosion in Natural Environments', *ASTM-STP* **558**, 352 (1974).
97. G. K. Berukschtis, G. B. Klark; *Korrosionnaja ustoitschiwost metallov i metallitscheskich pokrytij v atmosfernych uslovijach (Corrosion Resistance of Metals and Metallic Coatings under Atmospheric Conditions)*, Nauka, Moscow, 1971, pp. 94 to 115.
98. Y. N. Mikhailovskii, G. B. Klark, L. A. Shuvakina, A. P. Sanko, Y. P. Gladkikh, V. V. Agafonov, *Zaschtsch. Metallov*, **7**, 534 (1971).
99. K. Bartoň, D. Knotková, J. Spanilý, *Koroze a ochrana materiálu (Corrosion and Materials Protection)*, **17**, 85 (1973).
100. K. Bartoň, unpublished data.

Chapter 4

1. H. Kaesche: *Die Korrosion der Metalle (Corrosion of Metals)*, Springer Verlag, Berlin, 1966, Chapter 9: 'Pitting corrosion of passive metals'.
2. J. L. Rosenfeld: *Korrosija i zaschtschita metallov (Corrosion and Corrosion Protection)*, Metallurgia, Moscow, 1969, Chapter 3: 'Pitting corrosion of stainless steels'.

3. D. O. Sprowls, H. R. Brown, *Fundamental Aspects of Stress Corrosion Cracking*, NACE, Houston, 1969, pp. 466 to 509.
4. G. V. Akimov, N. D. Tomashov, *Korros. u. Metallschutz*, **8**, 197 (1932); **13**, 114 (1937); **15**, 157 (1939).
5. W. W. Binger, E. H. Hollingsworth, D. O. Sprowls: *Aluminium* (ed. K. R. van Horn), American Society for Metals, 1967, pp. 209–276.
6. E. A. G. Liddiard, W. A. Bell, *J. Inst. Metals*, **81**, 269 (1952).
7. R. W. Staehle, *Fundamental Aspects of Stress Corrosion Cracking*, NACE, Houston, 1969, pp. 3 to 14.
8. H. H. Uhlig, *Fundamental Aspects of Stress Corrosion Cracking*, NACE Houston, 1969, pp. 86 to 97.
9. D. A. Vermilyea, *Fundamental Aspects of Stress Corrosion Cracking*, NACE, Houston, 1969, pp. 15 to 31.
10. T. P. Hoar, *Fundamental Aspects of Stress Corrosion Cracking*, NACE, Houston, 1969, pp. 98 to 103.
11. E. H. Phelps, *Fundamental Aspects of Stress Corrosion Cracking*, NACE, Houston, 1969, pp. 398 to 410.
12. E. N. Pugh, I. V. Craig, I. A. Sedriks, *Fundamental Aspects of Stress Corrosion Cracking*, NACE, Houston, 1969, pp. 118 to 158.
13. E. Mattson, *Electrochim Acta*, **3**, 279 (1961).

Chapter 5

1. K. Bartoň, *Technische Rundschau.*, **54**, 19 (1970).
2. T. Väland, *Corrosion Science*, **9**, 577 (1969).
3. G. H. Cartledge, *British Corros. J.*, **1**, 293 (1966).
4. D. M. Brasher, A. D. Mercer, *British Corros. J.*, **3**, 120 (1968).
5. D. M. Brasher, *British Corros. J.*, **2**, 95, 122 (1967).
6. R. Bartoníček, L. Červený, *Chem. prumysl*, **8**, 622 (1958).
7. T. E. Evans, *Werkstoffe und Korrosion*, **15**, 797 (1964).
8. F. Eisenstecken, W. Stinnes, *Archiv fur das Eisenhuttenwesen*, **7**, 469 (1956).
9. K. A. Chandler, M. B. Kilcullen, *British Corros. J.*, **5**, 24 (1970).
10. D. Kuchyňka, unpublished results, Prague.
11. K. Bartoň, Ž. Bartoňová, *Werkst. Korros.*, **21**, 25 (1970).
12. M. Pourbaix, A. Pourbaix, *Rev. Mens. Soc. Royale Belge*, 195 (1968).
13. N. D. Tomashov, A. A. Lokotilov: *Korrosija i zaschtschita stalei (Corrosion and Protection of Steel)*, MASGIZ, Moscow, 1959, pp. 171, 178
14. V. V. Skorchelletti, S. E. Tukachinskii, *Zh. prikl. chimii*, **26**, 30 (1953); **28**, 651 (1955).
15. J. C. Hudson, J. F. Stanners, *J. Iron Steel Inst.*, **180**, 271 (1955).
16. G. Schikorr, *Metall.*, **15**, 961 (1961).
17. U. R. Evans, *Nature*, **206**, 980 (1965).
18. K. Bartoň, V. Veselý, *Strojírenství*, **13**, 980 (1963).
19. D. Čermáková, J. Vlčková, *Proc. 3rd Int. Congr. Metallic Corrosion, Moscow*, 1966, Vol. 4, p. 497 (English edition).
20. T. Biestek, *Metal Finishing*, **4**, 48 (1970).
21. H. Laub, *Metall.*, **19**, 830 (1965).
22. G. Schikorr, *Aluminium*, **43**, 108 (1967).
23. K. Bartoň, E. Beránek, *Werkst. Korros.*, **10**, 377 (1959).
24. K. Müller, *Werkst. Korros.*, **15**, 533 (1964).
25. J. Elze, *Metalloberfläche*, **22**, 97 (1968).
26. W. Friehe, B. Meuthen, W. Schwenk, *Stahl und Eisen*, **88**, 477 (1968).
27. F. Cooke, J. K. Cosslet, R. W. Scott, C. E. A. Shanahan, *British Corros. J.*, **1**, 283 (1966).
28. J. F. Stanners, *Protecting Steel by Sprayed Metal Coatings*, The Industrial Finishes Conference Convention, London, 1961.
29. J. Moravec, A. Chlübna, *Koroze a ochrana materiálu*, **12**, 89 (1968).
30. I. K. Wirth, W. Machu, *Die Korrosion des Eisens unter Schutzfilmen, insbesondere Farbanstrichen (Corrosion of Iron Under Protective Films, Especially Paint)*, Verlag Technik, Berlin, 1952.

31. E. Manegold, *Fette und Seifen*, 581, 675, 725 (1952).
32. M. Svoboda, D. Kuchyňka, B. Knápek, *Farbe und Lacke*, **77**, 11 (1971).
33. K. Schwabe, *Werkst. Korros.*, **18**, 961 (1967).
34. J. E. O. Mayne, *British Corros. J.*, **5**, 106 (1970).
35. U. R. Evans, C. A. J. Taylor, *Trans. Inst. Metal Finishing*, **39**, 188 (1962); **43**, 169 (1965).
36. J. Barraclough, J. B. Harrison, *J. Oil Colour Chem. Assoc.*, **48**, 341 (1965).
37. G. H. Cartledge, *British Corros. J.*, **1**, 293 (1966).
38. J. d'Ans, H. J. Schuster, *Farbe und Lacke*, **61**, 453 (1955).
39. G. Meyer, *Schweizer Archiv.*, 160 (1966).
40. J. E. O. Mayne, *J. Applied Chem.*, **9**, 673 (1959).
41. K. Bartoň, J. Hron, Ž. Bartoňová, *Coll. Czech. Chem. Commun.*, **25**, 585 (1960).
42. K. Bartoň, D. Kuchyňka, E. Beránek, unpublished data.
43. V. Čupr, B. Cibulka, *Deutsche Farbenzeitschrift*, 442, 476 (1964).
44. V. Čupr, J. B. Pelikán, *Metalloberfläche*, **20**, 470 (1966).
45. R. Bartoníček, *Coll. Czech. Chem. Commun.*, **23**, 1174 (1958).
46. L. Červený, J. Nemcová: *Inhibitory koroze kovů (Corrosion Inhibitors)*, SNTL, Prague, 1964, pp. 120 to 123.
47. S. Karabiberov: *Czech. Patent 139615.*
48. S. Karabiberov: Internal (unpublished) report, SVÚOM, Prague, 1968.
49. S. Karabiberov, K. Bartoň: *Czech. Patent 139615.*
50. J. E. O. Mayne, *J. Iron Steel Inst.*, **164**, 289 (1950).

Chapter 6

1. K. Bartoň, Ž. Bartoňová, *Werkstoffe und Korrosion*, **20**, 216 (1969).
2. J. C. Hudson, *Werkst. Korros.*, **13**, 363 (1964).
3. K. Bartoň, Ž. Bartoňová *Werkst. Korros.*, **21**, 85 (1970).
4. E. Längle, *Schweizer Archiv.*, **34**, 109 (1968); **34**, 147 (1968).
5. N. D. Tomashov, A. A. Lokotilov, *Zav. lab.*, **24**, 157 (1960).
6. P. J. Sereda, *Materials Research and Standards*, 719 (1961).
7. H. Guttman: *ASTM Special Technical Publication*, **435**, 223 (1968).
8. V. Marek, unpublished data, SVÚOM, Prague, 1969.
9. J. Honzák, unpublished data, SVÚOM, Prague, 1970.
10. J. A. von Frauenhofer, G. A. Pickup, *Anti-Corrosion*, **14**(3), 8; **14**(9), 11 (1967); **15**(2), 6 (1968).
11. H. Schwarz, *Werkst. Korros.*, **16**, 93, 208 (1965).
12. I. Kokoška, Internal report, SVÚOM, Prague, 1968.
13. D. M. Brasher, T. J. Nurse, *J. Appl. Chem.*, **9**, 96 (1959).
14. M. Bureau, *Corrosion-Traitments-Protection-Finition*, **16**, 235 (1968).
15. W. H. Ailor, M. R. Hodgson: *Punch Card Handling of Atmospheric Test Data*, Reynolds Metals Co., Richmond, U.S.A., April 1960.
16. J. Honzák, *Chem. listy*, **62**, 458 (1968).
17. A. G. Huber, 'Aerosol' corrosion test chamber, Zurich.
18. M. Rychtera, B. Bartáková, *Tropicproofing Electrical Equipment*, Part II, 'Climatic Tests', Hill (London) and SNTL (Prague), 1963.
19. K. Bartoň, *Werkst. Korros.*, **9**, 547 (1958).
20. K. Bartoň, E. Beránek, *Werkst. Korros.*, **10**, 377 (1959); **11**, 348 (1960).
21. K. Bartoň, *Proc. 1st Int. Congr. Metallic Corrosion*, Butterworths, London, 1962, p. 685.
22. K. Bartoň, Ž. Bartoňová, *Proc. 3rd Int. Congr. Metallic Corrosion*, Moscow, 1966, Vol. 4, p. 483 (English edition).
23. F. L. LaQue, *46th Annual Technical Proceedings American Electroplaters Soc.*, 1959.
24. L. Saur, R. P. Baco, *Plating*, **39**, 321 (1966).
25. M. Pražák, V. Spanilý; Tagungsbuch der 42. Veranstaltung der Europ. Federation Korrosion-Euromeskor, Prague, 1968, Vol. 3, p. 1.
26. G. Schikorr, *Werkst. Korros.*, **15**, 537 (1967).
27. H. Spähn, *Metalloberfläche*, **20**, 151 (1966).

Chapter 7

1. J. L. Rosenfeld: *Korrosija i zaschtschita metallov (Corrosion and protection of metals)*, Metallurgia, Moscow, 1970, Chapter 4: 'Corrosion and construction concepts'.
2. K. A. van Oeteren: *Konstruktion und Korrosionschutz (Construction and Corrosion Protection)*, Curt R. Vincentz Verlag, Hanover, 1967, Chapters 8 and 9.
3. *Korozní odolnost materiálů a povlaků (Corrosion Resistance of Metals and Coatings)*, SVÚOM, Prague, 1968, Volume 1.

Index

194